ROUTLEDGE LIBRARY EDTIONS: GLOBAL TRANSPORT PLANNING

Volume 3

SOVIET AND EAST EUROPEAN TRANSPORT PROBLEMS

SOVIET AND EAST EUROPEAN TRANSPORT PROBLEMS

Edited by
JOHN AMBLER, DENIS J. B. SHAW AND
LESLIE SYMONS

Routledge
Taylor & Francis Group

LONDON AND NEW YORK

First published in 1985 by St Martin's Press

This edition first published in 2021
by Routledge
2 Park Square, Milton Park, Abingdon, Oxon OX14 4RN

and by Routledge
605 Third Avenue, New York, NY 10017

Routledge is an imprint of the Taylor & Francis Group, an informa business

© 1985 John Ambler, Denis J. B. Shaw and Leslie Symons

British Library Cataloguing in Publication Data
A catalogue record for this book is available from the British Library

ISBN 13: 978-0-367-69870-6 (Set)
ISBN 13: 978-0-367-72606-5 (Volume 3) (hbk)
ISBN 13: 978-0-36772-609-6 (Volume 3) (pbk)

Publisher's Note
The publisher has gone to great lengths to ensure the quality of this reprint but points out that some imperfections in the original copies may be apparent.

Disclaimer
The publisher has made every effort to trace copyright holders and would welcome correspondence from those they have been unable to trace.

Soviet and East European Transport Problems

Edited by

John Ambler,
Denis J.B. Shaw & Leslie Symons

Library of Congress Cataloging in Publication Data
Main entry under title:

Soviet and East European transport problems.
    Revised papers from a colloquium held in Nov. 1983
at Greynog Hall, University of Wales.
    Includes index.
    1. Transportation—Soviet Union—Congresses.
2. Transportation—Europe, Eastern—Congresses.
I. Ambler, John, 1932-   II. Shaw, Denis J.B.
III. Symons, Leslie.
HE255.S63   1985   380.5'0947   85–11917
ISBN 0-312-74757-8

CONTENTS

LIST OF TABLES

LIST OF FIGURES

PREFACE

John Ambler, Denis J. B. Shaw and Leslie Symons

This book contains the revised versions of papers
delivered in November 1983 at the second Gregynog
colloquium on Soviet and East European transport
problems. Participants came from a variety of
disciplines, such as geography, economics and
political science, and included specialists in
transport with experience of working in the various
industries. Although many of the contributors were
from the U.K., the United States, West Germany and
Sweden were also represented. This colloquium was
jointly organised by Denis Shaw and Leslie Symons,
and funding was generously provided by the Ford
Foundation (through a grant administered by the
National Association for Soviet and East European
Studies and the British Universities Association of
Slavists), the British Council and University
College Swansea's Centre for Russian and East
European Studies.
    As on the first occasion, the central focus of
the second colloquium concerned the way in which
transport functions and is organised under the
centrally planned economic system. However, in the
eleven years since the first meeting, the world
economy has suffered the effects of recession, and
the communist states have been unable to escape its
repercussions. The poor performance of the transport
sector in many of the centrally planned economies
has been blamed for contributing to a disappointing
economic record. It was therefore inevitable that
the participants at Gregynog were interested not
only in the organisation and problems of transport
for their own sake, but also in the wider
implications of transport difficulties for the
various communist economies and their future
development. Such an emphasis is especially
important in view of a lack of attention to

communist transport problems in the economic
literature generally.

A perusal of the chapters of this book will
quickly lead to the realisation that the
transportation systems of the various states have
much in common. It will be useful to say something
about these common features at the outset. In the
first place, the systems all exist within the
constraints of the centrally planned economy with
its characteristic peculiarities in terms of the
central allocation of investment funds, administered
prices, directive planning, and a marked degree of
political control. Traditionally, transportation
investment has tended to lag behind demand,
producing a continual drive to release 'inner
reserves' within the sector. As with other sectors,
the emphasis has tended to be upon quantity rather
than quality of service. However, new attitudes now
seem to be germinating : for one, the accent on
tonnage moved may be giving way to a concern with
such qualitative criteria as utility and speed. The
years ahead will reveal how far these new approaches
will be allowed to develop, though it is already
apparent that vested interests and entrenched
outlooks will in fact be difficult to dislodge.

A second feature common to the Soviet and East
European transport systems, and also quite at odds
with West European and American experience, is the
dominant position of the railway. To some extent
this follows from the low priority given to
transport when it comes to investment and the
failure, until relatively recently, to develop
alternative modes. The successful release of the
inner reserves of the communist railway systems,
including that of China, enables them to carry three
quarters of the world's rail freight traffic today,
compared with 40% in the early 1950s. By comparison
roads have been greatly neglected, though both road
and pipeline networks are now growing very quickly,
starting from a small base.

A further common feature derives from the fact
that all the communist states have striven, to a
greater or lesser extent, to base their development
wherever possible upon their own resources and to
avoid overdependence upon the outside world,
especially the capitalist world. This autarkic
tendency is most marked in the case of Albania, and
even Yugoslavia has not escaped it entirely.
Communist transport systems are affected
accordingly. The determination to develop national
resources fosters interregional communications

within each country, although the regional resource imbalances which exist in several of them produce severe transport difficulties. The heavy emphasis placed on fuel, steel and construction means that, with the exception of sand and gravel, the commodities concerned have their sources in few places but are sent to many. Siberia, with its territorial production complexes, has become more and more important as a treasure house of the Soviet economy despite its hostile environment. Transport has to make its due contribution to the exploitation of these resources. Unfortunately for the rail network, this is also the region where transport already experiences the greatest strain. Similar difficulties arise in Poland, especially in Upper Silesia.

Difficulties of another kind concern the question of rural development. None of the communist states can afford to neglect agriculture and to become too dependent on the outside world for the supply of food. Unfortunately, rural backwardness is an acute problem in several of these countries. Rural development policies are therefore accorded a certain emphasis, but such policies can also place strains on the transport network, and especially on the poorly developed roads.

While autarkic tendencies have thus had a certain influence on the transport systems of the communist states, the time has long since passed when these states could cut themselves off completely from the outside world. New patterns of trade and contact between the communist countries, and between each of them and the wider world, now also affect their transport networks and sometimes add to their problems. Several of the chapters of this book touch on these international aspects, while that on Soviet shipping emphasizes the global dimension whose significance has now become quite marked.

In this preface we can only point to some of the most obvious of the common features and connected problems which underlie the chapters of this book. Perhaps more intriguing are the questions raised in several of the chapters concerning the purpose and the utility of transport, and how to measure its efficiency. Thus many of the contributors note that Western governments speak with forked tongues about transport policy. The latter say that transport prices must reflect opportunity costs, and then intervene to prevent this. For instance, Western transport agencies, and also Western scholars, tend to regard the flexibility that comes from the provision of spare capacity in transport as a positive advantage, even though it has a real cost. Therefore the fact that the average Soviet freight wagon moves two and a half times the daily distance of one in the USA might be regarded by them as ambivalent evidence of efficiency. Western opinion would tend to see advantages in a transport system with more flexibility than the Soviet one appears to have. It would certainly look askance at the Soviet level of occupancy of rail passenger carriages, which is roughly double that in West Germany and neighbouring countries.

Double standards concerning efficiency in transport also afflict the centrally planned economies. The officially approved policy to reduce transport costs to a minimum is sometimes tempered, as noted already, by a concern for quality. On the other hand, the costs of transport are a major burden under any political and economic system, since transport requires an expensive infrastructure of routes and terminals (such as stations and ports) before the individual vehicles can even move. The control of such costs plays an important role in the communist economies with their traditional emphases upon producer goods and the military. Several contributors suggest that such frugality with resources has now been pushed too far and that past underinvestment has sown serious problems for the future. In the meantime the campaign to release yet more inner reserves in the system continues. Thus far the campaign has had limited success, although the reasons for this appear to be complex. Several of the chapters note organisational and managerial inefficiency, lack of effective computer applications, shortages of spares, poor labour morale, and a sub-optimal mix of vehicles. It is difficult to rank the various factors at work. There is, however, general agreement that the centrally planned economies are not as successful as Western economies when it comes to innovation. In transport the obvious example is containerisation which has proved slow in terms of development, though there are many other problem areas as well. The bureaucratic difficulties associated with horizontal linkages (deriving from the 'narrow departmentalism' noted in several papers) are absent in Western economies and this means that Western design time is generally shorter.

Perhaps the most obvious conclusion to draw

from all this is the fact that, in both East and West, transport serves competing economic and social goals and, in comparing systems, much depends upon the values of the observer. Judgements are also complicated by our ignorance or uncertainty concerning many policy objectives. In the case of Soviet railways, for example, electrification has already brought vast improvements in efficiency and performance. But it is unclear whether the primary aims of policy in this sector are to release labour to manufacturing, to get more output per vehicle, to increase the profit per unit of cost input, to reduce energy consumption, or some combination of these. Likewise the aims of Soviet shipping policy could be to get hold of hard currency at almost any price, to weaken Western shipping, to offer a cheap service to the Third World, or to further military objectives. The authors of these chapters will be satisfied if they succeed in casting some light on the many issues involved.

A brief synopsis of the chapters that follow will help to illustrate some of the above matters in their context. The first chapter, by Denis Shaw, covers regional and branch (or ministerial) planning in the USSR. It sets the scene for the specialised chapters which follow by describing the main themes in detail and examining how they criss-cross and conflict with one another. Recent policy developments to encourage inter-branch co-ordination are examined and all are found to be to some extent wanting, in the absence of more fundamental reforms.

Chapter 2, by John Ambler, Holland Hunter, and John Westwood, starts with the fact that since Soviet railways carry more than half the world's rail freight, this alone puts them in a league of their own. Post-war growth until the late-70s was large and consistent, and they document how the release of 'inner reserves' (in particular, the switch to electric and diesel traction, and 4-axle wagons) made its contribution. The recent plateau in performance, seen with horror by Gosplan, indicates some deep crisis. Gosplan's solution that improved management is the dynamo generating greater efficiency is shown to face several stumbling blocks. The declines in asset utilisation, and in railway profitability, coincide with and reinforce lower morale; significantly, the gap between Soviet and US performance, static for so many years, has shown signs of increasing. All in all, the railways are in a fragile condition, years of limited investment and heavy utilisation having brought them nearer to the point where intense usage results in catastrophic breakdown. However, the recovery of 1983 and 1984 shows that it needs a bold prophet to forecast imminent collapse; each fresh challenge seems to find an answer from deep within those 'inner reserves'.

David Wilson in chapter 3 surveys the key position of energy in the Soviet economy and its role in transport demand. Energy is a universal requirement, nationally and internationally, with sources in ever more difficult and inaccessible regions, appearing in a variety of forms which are to some extent interchangeable, and transported in bulk in four totally dissimilar ways : by pipelines, by tanker, by tank car, and by electricity grid. The interplay between modal choices is partly determined by product type and market, and its cost at pithead and wellhead. The chapter concludes by examining the Long Term Energy Programme in the light of these factors in order to predict future modal flows of each energy type.

Shipping is the transport mode par excellence for international movements. Paul Lydolph in chapter 4 shows in how many ways the Soviet intervention has been unwelcome to the West. The underlying ironies are exposed : the alleged quasi-monopoly of the conference system (the Third World view); the avoidance of controls by registration in 'flag of convenience' countries (the developed world's earlier worry); the slow acceptance of containerisation in the communist world; the general criticism of trade carried in third country ships, except of course by the third countries themselves. The Soviets have moved deftly into a market long dominated by British shipping; by creaming off liner trade, by offering a cheap and adequate Trans-Siberian land bridge for containers, and by supplying cruise liners for a near-luxury market.

In chapter 5, Leslie Symons discusses Soviet air transport, a clear case where military pressures obscure the statistics. Nevertheless, the USSR has been falling behind the USA in modern aircraft of the jumbo and airbus type, something obvious to even the most casual observer. Better hidden, but of great interest in the present context, are the organisational problems on which limited information is obtained from the specialist press with its publication of criticism by staff and journalists.

The final chapter in this group on Soviet affairs, by Martin Crouch, covers the 'poor relation', road transport. He makes a significant

point which we in the West tend to forget : the
Soviets refer to road transport as 'motor
transport', as if to underline the generic problem
of roadlessness. The problem of the efficient use of
trucks owned by production enterprises, and the way
in which bureaucracy exercises its all-interfering
control, are delineated in a series of examples.
Road passenger transport by different modes is also
covered in some detail. Chapter 6 concludes that
this form of transport, despite all its problems,
has one important effect : it enables the 'second
economy' to work.

For the German Democratic Republic, Johannes
Tismer identifies a major official objective as
reducing the energy requirements of transport. This
deceptively simple aim is shown to have numerous
side effects and spin-offs. His analysis of the
development of truck fleets controlled by individual
enterprises, the better to meet their production
plans, is positively Galbraithian in exposing the
Marxist equivalent of 'public squalor, private
wealth'. The proposed tactics will, he believes,
result in more bureaucratisation because of the
extra controls needed to harmonise movements and
production; and more constraints on management,
because of the policy on regional specialisation and
the disturbance to established trading patterns. The
attempts to impose unreal plan targets are also
unlikely to be effective. Even in the GDR few steel
managers will worry about how many ton-kilometres a
ton of steel from their plants will generate.

The Albanian railway system moves as much
freight in a year as the Soviet system does in under
an hour. In chapter 8 Derek Hall describes the
political and ethnic difficulties in the
international linking of Albania, hitherto unique in
Europe in having a railway with no cross-border
connection with any other country. Fate has dealt
Albania a poor hand in that the only feasible link
at present is with Yugoslavia, a country itself
beset with ethnic problems and already suffering
from a relatively large (12%) share of transit
traffic. First hand knowledge of Albanian experience
is very scarce; despite the relative insignificance
of Albania on the world scene, Hall's insights into
this long forgotten and isolated ideological
backwater are welcome, and raise many issues which
have counterparts in other chapters.

The final chapter, by Claes Alvstam, Zygmunt
Berman and Andrew Dawson, covers both the Polish
internal transport situation and the flows to and

from the USSR. Poland accounted for one sixth of all European freight rail traffic[1] in 1981, nearly one and a half times as much as Romania, the next largest country. Total freight transport in 1982 was almost a fifth below that of 1980, and the still relatively abundant statistics enable the problems to be charted with reasonable accuracy. The chapter analyses the pressures imposed by the Soviet Union, together with the long distances over which Soviet exports travel, the rising opportunity costs of such freight, and the peculiarities of the tariff. In a sense, it can be said to illustrate one important issue : the essential interdependence of individual transport systems, and their interrelationships with the national economy and with the economies of neighbouring countries.

The papers read at the November 1983 colloquium stimulated a good deal of discussion and debate and the participants felt moved to plan for a further gathering within eighteen months' time. This is to be held in Berlin, and it is hoped that there the many points of disagreement and areas of ignorance will be subjected to further scrutiny. The aim of this and possible future meetings is to provide a continuing forum in which the specialists in the various fields can exchange views and stimulate analysis in this neglected and yet challenging field of study.

[1] i.e. outside the USSR.

ACKNOWLEDGMENTS

The contributors wish to express their gratitude to the National Association for Soviet and East European Studies and the British Universities Association of Slavists for their support of the Gregynog symposium by means of a Ford Foundation grant. Thanks are also due to the British Council for funding participation by overseas visitors. The University of Wales kindly provided facilities and transport for the occasion.

TRANSLITERATION AND MEASUREMENTS

The System of Transliteration used is that of the American Board on Geographic Names. Metric measures are exclusively used, with the exception of the chapter on shipping. Tons are metric tons.

SOVIET AND EAST EUROPEAN TRANSPORT PROBLEMS

Chapter 1

BRANCH AND REGIONAL PROBLEMS IN SOVIET
TRANSPORTATION

Denis J.B. Shaw

Since the 1960s the USSR has been attempting to
pursue an economic development policy which
emphasizes the need for an efficient use of inputs.
In an earlier period of industrialisation, when
labour and raw materials were relatively abundant
and returns to investment were high, the economy was
able to grow by extensive means. Nowadays, with
shortages of energy and other raw materials in the
most settled and developed part of the country and
with decreasing rates of growth in the labour force,
the accent is on seeking economic growth by means of
new developments within the production process
itself. In this situation the transportation system
is seen as having an important part to play.
Numerous policy statements have stressed the
necessity for a more intensive use of the system,
for the removal of bottlenecks and for a timely and
effective investment policy, consistent with the
needs of the economy as a whole. The goals of
transport under the intensification strategy are
neatly summarised in the opening words of a recent
publication : 'The country's transport system is a
highly intricate complex, intimately interconnected
with all branches of the economy, the major task of
which is the delivery of all products to the point
of consumption for the minimum outlay in resources
and time' (Kolesov, 1982, p.3).
    In the most recent period the pressure to
improve the performance of the transportation sector
has if anything grown. The country's disappointing
economic record, especially since the latter half of
the 1970s, has goaded its political leaders into an
ever more frantic search for economies and
under-utilised reserves. Transport is now seen as a
definite barrier to economic growth. Serious
complaints about its performance began to be voiced

1

in the late 1970s. In his report to the 26th
Congress of the CPSU in March 1981, party General
Secretary V.I. Brezhnev listed a number of serious
problems which would afflict the performance of the
Soviet economy in the 1980s. Prominent among them
were the inadequacies of the transportation system
and the poor state of the roads. [1]
    Investment in transportation has rarely been
the highest priority for the Soviet leadership. As
the economy has grown, so the demands made upon the
system have multiplied and efficiency has often
suffered. Problems in planning, which have also
afflicted other areas of the economy, have only
exacerbated the situation. But transport's
difficulties are not merely the consequence of
inadequate investment and organisation. Recent years
have witnessed a whole series of new developments in
the economy which have greatly increased the strain.
Foremost of these, and listed by Brezhnev among his
major problems of the 1980s, is the development of
new resource bases in the remote and difficult
territories of the east and the north. The
development of these is already placing great
strains on the transport system. At the present
time, virtually all the increases in the production
of oil and natural gas and 90% of those in coal
come from Siberia (Kolesov, 1982, p.168). In 1979,
oil and oil products accounted for 21.8% of the
freight shifted by the transportation agencies, and
coal 14.4% (Kolesov, 1982, p.78). [2]   Siberia   is
the USSR's main base for increases in timber
production. Outlying mineral workings in the remoter
parts of Siberia and the Far East are also making
greater demands on the transport system, and these
demands can only increase in the future.
    Other factors which have complicated
transport's performance in recent years include the
USSR's international links, and the need to improve
agriculture's performance by developing the rural
infrastructure and placing processing industries in
the countryside. A characteristic feature of the
system is the growing interconnectedness between
regions in terms of freight flows. One study of
interregional freight flows over a 17-year period
(1960-77) showed the most rapidly developing flows
to be those between the most distant regions - i.e.
between the north, west and south of European USSR
on the one hand and the eastern regions of Siberia
and the Far East on the other (Kolesov, 1982,
pp.94-5). The considerable increases in lengths of
haul must also be noted in this connection : average

2

length of haul by rail grew from 798 km in 1960 to
908 km in 1979, on all forms of transport from 175
km to 212 km and on all forms apart from sea and
road transport, from 744 km to 977 km (Kolesov,
1982, p.62). These are among the longest lengths
of haul in the world. The pressures on the system
are therefore great, and some portions of it are
particularly oversubscribed. For example, Siberia's
rail and hard-surface road system has an average
freight density some 5-6 times that in the Baltic
republics. Pressure on Siberian railways is greater
than anywhere else in the country on average and is
particularly significant· in West Siberia where
freight turnover per kilometre of route is twice
that for the USSR as a whole, and in some places
exceeds the USSR average by a factor of five or six
(Kolesov, 1982, p.177). Unfortunately, these
pressures have tended to increase over time and have
no doubt added to the problems which the transport
system now faces.

The Structure of Transport and Transportation Planning

A point regularly made in the Soviet literature on
transportation is that the USSR possesses a 'unified
transport system'.[3] This means that the entire
system of transport - roads, railways, waterways,
pipelines and so on - is owned by the state. It can
therefore be planned as a unity, with all parts
complementing and co-operating with one another,
thus eliminating waste and unnecessary competition.
The claim is also made that this permits the
optimisation of the transport system. This is
because not only that system but also the entire
economy are under central control, and thus all the
necessary linkages can be planned from the centre
and future changes anticipated.

Needless to say, such claims are only very
partially substantiated by the facts. Enough has
been written concerning the Soviet economy to
indicate the extreme difficulty of planning such an
enormous machine and ensuring the necessary degree
of co-ordination between its many parts (e.g. Nove,
1980). Claims regarding optimisation have to be
treated with a good measure of scepticism. Western
economic analyses have indicated that the very
structure of the Soviet economy, based as it is upon
branch administrations operating from the centre, is
one of the numerous factors which make life
difficult for the planners. Thus in transportation

3

the various transport modes - rail, inland waterway, shipping, airways and so on - are controlled by different branch ministries and, as is common in all parts of the Soviet economy, co-operation between branches is often difficult to secure. Under the prevailing system, vertical rather than horizontal linkages predominate and the natural tendency is to ignore or to minimise possible interconnections with other branches. The charge of departmentalism (vedomstvennost) - of forgetting horizontal linkages and of putting the interests of one's own department first - is frequently levied against the transportation agencies and cited as one of the factors militating against optimisation in the system.

Defects of the branch approach are all-pervasive and have many different manifestations. One frequent complaint is that while parts of the transportation system are working beyond capacity, other parts which could provide alternatives are underutilised. The most commonly quoted example is the underuse of waterways compared with the more expensive and oversubscribed railways, a phenomenon which arises mainly because branch ministries and their enterprises prefer to use the more reliable and generally speedier railways. One writer cites a particularly notable case of this in the shipping of timber along the heavily overburdened Trans-Siberian railway between Novosibirsk and Omsk in West Siberia (Kolesov, 1982, p.195). Presently timber from the area of Kolpashevo on the Ob' is mainly transported upriver to Novosibirsk for transshipment to rail. However, in view of the heavy pressure on the West Siberian railways, it has been shown to be economic to send the timber downstream along the Ob' to its confluence with the Irtysh, and then upstream along that river to Pavlodar and Semipalatinsk for onward delivery by rail to points in Kazakhstan and Central Asia. Timber could also continue to be sent upstream along the Ob' to consumers in Tomsk, Novosibirsk and Barnaul. Despite this solution to the West Siberian rail problem, however, there are a number of departmental barriers. Firstly there is the reluctance of the Ministry of the Timber Industry, for already-cited reasons, to use the waterways. Secondly, and perhaps even more surprising, there is the failure to co-ordinate the activities of the different waterway authorities. Thus there are three navigation authorities on the Ob'-Irtysh system who fail to agree among themselves. For example the West

Siberian navigation authority forbids the transport of timber on rivers in Tomsk region, in the basin of the Ob', in craft belonging to the Irtysh authority.

Problems such as these are caused not merely by bureaucratic ineptitude. The planning system itself does much to encourage such tendencies. It is a characteristic of Soviet planning that, although major decisions are made by planners at the centre, many day-to-day decisions, even including quite important ones, are left to the branch ministries to sort out. This occurs not merely because of the complexity of the system, but also because branch ministries and their organisations are frequently in the best position to take decisions of a technical or local nature. The State Planning Committee, Gosplan, is primarily concerned with inter-branch issues, though the committee does have separate departments to deal with major sections of the economy and this itself can reinforce branch tendencies. Transportation ministries like others are hard-pressed to fulfil their plans and meet the many demands made upon them. In this situation there is little time or incentive for them to seek out economies through innovation or through new forms of co-operation with other branches. The complex and hierarchical process of decision-making discourages the latter in any case. Of course it should be the work of the central planners to take a long-term view and to seek out economies by fostering inter-branch co-ordination. In practice, however, this is difficult to achieve. Transport is planned over the long-term (for periods of five years or more) in connection with the planning of the economy as a whole. Thus, from forecasts of economic performance over a five-year period, transportation needs can also be projected based upon past experience. However, planning even over five-year periods is fraught with difficulties and five-year plans usually have to be revised on a regular basis. As far as annual planning is concerned, planners both in Gosplan and the transportation ministries rely upon such methods as proportions of dispatching and of regional balance but these are vague and give only a general picture of transport requirements (Mieczkowski, 1978, pp.51-67). In practice, therefore, great reliance is placed on a system of contracts whereby enterprises and transportation agencies enter into agreements before a plan period which express the service to be provided during that period. Although such contracts are approved by Gosplan at all-Union or republican

5

level, it is a relatively <u>ad hoc</u> arrangement which
may or may not correspond to the annual plan. Indeed
one Western investigator went so far as to deny the
practical importance of planning in East European
transportation.[4]

In arranging their own dispositions based upon
annual plans and contracts, the transportation
agencies rely upon a series of traffic flow norms
derived from past experience. However, it appears
that the Soviet norms have become out-dated and
certainly do not lead to an optimal use of the
transportation system (Shafirkin, 1975, p.49). The
July 1979 economic decree ordered their renewal.[5]
In the present situation, with transportation being
in effect a scarce commodity and transportation
agencies given plenty of scope for making their own
arrangements, departmentalism is encouraged. Thus
the agencies are inclined to give priority to
freights which are easier and cheaper to carry and
are able to choose routes which best serve their own
interests. Past reliance upon freight ton-km as a
major indicator of transportation performance has
encouraged over-intensive use of fixed capital and
increased length of haul, to the detriment of the
economy as a whole. Some of these problems could of
course be counteracted by greater use of economic
incentives and the introduction of market-type
prices. But the Soviet Union has so far eschewed
such radical measures. Although some use is made of
economic levers in transportation, administered
prices are by no means the same as scarcity prices.
The prevailing system is in effect one of rationing,
in which enterprises are inclined to exaggerate
their transportation needs, knowing that their
demands will inevitably be pared down, and in a
sellers' market transportation agencies are able
to order their affairs to suit themselves rather
than their clients.

Departmentalism also manifests itself in
investment policy. Current investment criteria tend
to favour reconstruction rather than investment in
new infrastructure and success indicators pay
premiums for increasing the intensity of use of
existing fixed capital rather than for making new
and timely investments. Transportation agencies are
therefore prone to be materially indifferent to the
further extension of their networks. In the case of
Siberia the railway system has thus tended to
develop in piecemeal fashion according to immediate
needs and to serve particular individual
developments. New industrial plants, mines and

timber enterprises have been connected by rail to
main lines with little account taken of the
long-term comprehensive development of new regions
and territories. As already noted, the overloading
of east-west lines across the West Siberian plain is
a particular problem which has arisen as a result of
this ad hoc approach. It is interesting to note that
sometimes the problems of the railways are
exacerbated by the independent activities of other
industrial ministries. Thus the Ministry of the
Timber Industry stands accused of building its own
timber-carrying railways often parallel to navigable
waterways in parts of West Siberia. However the
overpressurised main lines are unable to accept the
burden and timber stands in station yards and
warehouses for periods of up to five years (Kolesov,
1982, p.195).[6] Problems also arise as a result
of agreements between transportation and other
ministries. Thus, instead of being constructed along
a more northerly east-west route, the railway to
Surgut in north-west Siberia was built along a
shorter axis to connect with the already
overburdened junction at Sverdlovsk in the Urals
(Kolesov, 1982, p.184; North, 1979, pp.200-204). Ad
hoc development on a departmental basis therefore
gives rise to obvious bottlenecks which can have
negative consequences for the entire economy.[7] At
the present time, added investment is going into the
hard-pressed West Siberian railways but the accent
appears to be on encouraging further intensification
of use rather than on extending the network
(Biryukov, 1983, p.6).

Transportation and the Industrial Ministries
From what has been said so far, it is evident that
to characterise the Soviet transport network as a
'unified transport system' is to ascribe to it more
than is its due. Soviet transportation is afflicted
by departmentalism and as such finds it difficult to
contribute to the programme of economic
intensification. But problems of transportation
derive not merely from the structure of that system
in itself. They are also caused by the structure of
other parts of the economy and the way that they
impinge upon the workings of transport.
    One problem for Soviet investment policy that
has long been recognised is the difficulty, in the
absence of market-determined prices, of constructing
a method of comparing the effectiveness of
alternative investments. Thus, within the field of

transportation, comparisons between the economic effectiveness of alternative transport modes are fraught with difficulties (Kolesov, 1982, p.203; Mieczkowski, 1978, p.83). Similar problems arise when assessing the overall impact of particular investments. Spectacular losses of land, timber and minerals induced by hydro-electric power developments along major rivers have been encouraged by the failure to price natural resources. They have also been encouraged by the political weight of the Ministry of Power (Gustafson, 1981). Moreover, transport has suffered in consequence. Current methods of calculating investment costs fail sufficiently to take into account the losses which result from closing rivers to navigation. For this reason dams have been built on many rivers without locks to permit navigation in the belief that the economic consequences of this policy are negligible. Numerous writers argue that this is not so and blame faulty methods of measuring investment efficiency.

The above problem results from what is essentially a branch approach to the question of assessing investment. But the branch nature of the Soviet economy has numerous other implications for transport efficiency. One frequently quoted example comes from the field of road transport. In the USSR road transport plays second fiddle to railways especially for long-distance freight movements, and investment in roads at the local level is often dispersed among numerous agencies. The road system in consequence is poorly developed. Nevertheless, industrial ministries require road transport for short hauls. Calculations have indicated that the most economic way of arranging this would be to centralise road transport so that vehicles would be hired out to ministries and enterprises requiring their use. Such a procedure is said to produce significant economies in the use of vehicles and also manpower (Kolesov, 1982, pp.65-9; Shafirkin, 1975, p.26). Despite such savings, industrial ministries prefer to hang on to their fleets of vehicles and have successfully defied official policy declarations over the years. Of course from their point of view the attitude of the ministries is perfectly rational. Given the grave imperfections of the planning system, it makes sense to be in a position to ensure one's own transportation services and not to have to rely upon the uncertain performance of outside organisations. It is also true that similar tendencies are to be found in Western-type economies and transport costs are probably increased in consequence. Unfortunately, under the Soviet system, the economic costs of such autarkic practices are frequently not borne by the ministries or enterprises which adhere to them.

A further manifestation of ministerial autarky is to be found in the phenomenon of the cross-haul. Ministries try to combat the notorious inefficiencies of the supply system by building up their own subsidiary enterprises wherever possible. For this reason ministries have sometimes been described as 'multi-branch complexes', little empires which control a whole range of interdependent activities. Flows of goods belonging to these different empires pass one another along transportation routes, even though the goods may be identical. Upwards of 3-4% of the work of the transportation agencies is said to be wasted in this way (Shafirkin, 1975, p.69). A particular problem in this regard is the construction industry. Building materials, quite apart from timber, account for about a quarter of the freight shifted in the USSR on railways. In some areas they account for 35-40% (Mozhin, 1983, p.8). Length of haul naturally tends to be short, although it is not infrequently much longer than is considered optimal. In 1980 West Siberia imported about 41% of its requirements in building materials from other areas and, although there have been efforts to organise the manufacture of reinforced concrete materials in all regions, over 55% of production is transported across the boundaries of oblasts and krays. Mozhin blames the autarkic tendencies of branch ministries who create 'household economies' and supply themselves with their own requirements, regardless of transport consequences (Mozhin, 1983, p.8). Other bulky goods which are subject to much cross-hauling include timber and grain. However in this case an important contributing factor is the irrational location of industry. For example, a great deal of grain is shipped into Moscow for milling from eastern and south-eastern regions, only to be sent back as flour to the Transcaucasus, Kazakhstan and Central Asia. Similarly, the Soviet Far East imports roundwood by rail from Siberia, partly for export, but sends back timber to Siberia in the form of sawnwood, pit props and sleepers.

It is worth emphasizing at this point that the cross-hauling of goods, as well as such factors as the empty running of freight wagons and vehicles, are by no means characteristic only of Soviet-type economies. Such practices are extremely widespread

in Western economies, although the costs involved
are usually lost in company accounts. However, the
important point is the fact that a socialist
centrally-planned system ought in theory to be able
to minimise such tendencies. This is why the failure
to do so is so often a matter for official
complaint.

In his report to the 26th Congress of the
Communist Party in 1981, Brezhnev noted that : 'As
with many other problems, that of transport cannot
be solved in isolation. The reduction in outlays on
transportation is a great national objective. The
way to its achievement is by the rational location
of production, the development of optimal schemes of
freight shipment, the elimination of cross-hauls'
(Materialy, 1981, p.40). In mentioning the 'rational
location of production', Brezhnev drew attention to
one solution to the problems of transportation
systems which lies outside the sphere of
transportation itself. Undoubtedly many of the
problems of transport derive from the
departmentalism of the branch ministries in
questions relating to industrial location. Although
location of industrial plant in the USSR requires
official approval at national level, especially when
large investments or inter-branch co-operation are
involved, in practice a good deal of discretion is
left to the ministries. This occurs largely for
technical reasons. Naturally, ministries are subject
to official guidelines and to political and cost
considerations but it seems that even in the case of
quite large investments the approval of Gosplan and
other central agencies is often a mere formality
(Smolyar, 1976; Problemy, 1978). The result is that
locational decisons are frequently made without due
consideration for such externalities as the
availability of labour and costs of transportation,
to say nothing of housing, land and other factors
(Shaw, 1983). It is a common complaint of transport
economists that industrial location decisions are
often not made in the light of real transport costs
whose effectiveness is frequently diminished by
prevailing methods of planning and pricing
(Shafarkin, 1975, pp. 30-45; Dmitriyev, 1977,
pp.150-4). The overconcentration of industrial
capacity in certain regions or cities at the expense
of others is to the detriment of rational
transportation. It is encouraged by pricing policies
which permit ministries to be relatively indifferent
to external costs such as transport costs. Soviet
writers frequently castigate examples of irrational

location. Mozhin, for example cites the building of a paper and cellulose combine at Kzyl-Orda in Kazakhstan, supposedly based on local raw materials but in fact using non-local supplies of wood and cellulose. Similar irrationality is apparent in the case of a factory in Tallinn making excavators but based on non-local supplies of metal and fuel and sending more than 95% of its production to the southern or eastern regions. Apparently considerable economies could be derived from more sensible location and organisation of branches of the machine-building industry (Mozhin, 1983, p.8). But the problem with such examples is that they appear to be impressionistic and fail to analyse the detailed cost structures of the firms involved. [8]

A further interesting example of the problems which arise from a branch approach to industrial location comes from an analysis of the industrial supply system in the West Siberian oil and gas complex (Lebedeva, 1983). One of the effects of the 1965 economic reforms was to reorganise the state supply organisation known as Gossnab, concerned with the planning of industrial supplies passing primarily between one ministerial system and another. At the same time a number of branch ministries were permitted to retain their own supply organisations which were mainly concerned with intra-ministerial supplies. Needless to say, such a fine demarcation of duties does not always apply in practice and some duplication of effort does go on (Nove, 1980, p.41). This is particularly true in pioneering regions such as West Siberia. In this area Gossnab has been slow to organise its own network of supply bases, with the result that ministries and organisations are free to develop their own. In Tyumen' region alone some 55 supply organisations now operate under the administration of different ministries and departments, and each organisation often has several bases of varying sizes and of type of work. In some zones supply bases belonging to different ministries are placed side by side for many kilometres, each with their own warehouses, services, transport systems and so on. Where the bases are located along the banks of rivers, each has to have its own landing stage and equipment, but frequently boats cannot be unloaded at one pier because the equipment is already overloaded whereas a neighbouring pier under another ministry is free and its equipment stands idle. Similarly supplies cannot be switched between neighbouring bases under different ministries in

order to make up for temporary shortages because of
the usual bureaucratic barriers. Not only are
transport services hindered by such situations, but
there are also more far-reaching economic losses.
Centralisation of the whole supply system would
clearly solve some problems, but on the other hand
it is the lack of co-ordination in the supply system
to begin with which helps to encourage ministerial
autarky.

Transportation and Regional Co-ordination

The idea of fostering inter-branch co-ordination in
the Soviet economy is not a new one. It was after
all one of the factors which lay behind Khrushchev's
launching of his sovnarkhoz reform in 1957, and an
aim that was specifically mentioned by Brezhnev in
his address to the 23rd Party Congress in 1965
(Pavlenko, 1975, p.84). At the present time, as
shall be seen below, a number of practical
experiments have been implemented with this aim in
view, particularly at the regional level. Some of
these experiments will be reviewed with special
reference to their implications for transport.

In 1957, as just noted, Khrushchev launched his
sovnarkhoz reform. The essence of this reform was to
abolish the system of industrial branch ministries
and to replace them with over one hundred regional
economic councils, each responsible for organising
the industry in its own area. Part of the rationale
of the reform was to counteract the ministerial
autarky which had been prevalent under Stalin,
giving rise among other things to irrational
transportation. However, as is well known, the
reform was a failure and by the early 1960s
Khrushchev was forced to begin a covert
recentralisation. In 1965 the branch ministries were
restored, but attempts were made to obviate the
reappearance of departmentalism by retaining Gossnab
and by giving a greater role to various forms of
territorial planning (Pavlenko, 1975, pp.80 ff.).
These efforts did not produce the desired results.
In 1973 the government approved the association
reform, establishing larger groupings between
enterprises known as industrial and production
associations. At the time it was hoped that this
might permit the reorganisation of production and
thus the rationalisation of freight-flows within
associations. It was also hoped that associations
might arise which crossed ministerial boundaries.
Once again such hopes have been frustrated (CDSP
XXV no. 23 pp.1-4)

In spite of such frustrated efforts, the

Soviets are pressing ahead with other policies
designed to achieve similar effects within the
constraints of the economic system as presently
constructed. For one thing emphasis is now being
placed on long-term planning, including the
long-term spatial planning of industry and
settlement (Nekrasov, 1975, pp.52ff.). Work on the
so-called General Schemes for the distribution of
production and of settlement began in the 1960s, and
at present the third General Scheme for production
is being composed, to operate for a fifteen-year
period (Mozhin, 1983, p.10). It is hoped that by
taking such a long-term view, technical and economic
changes can be anticipated and investment can be
channelled through the branch system towards
approved goals. The General Schemes are being
augmented by more detailed locational schemes worked
out by branch and by region. At the regional level
co-ordinated development is to be secured by means
of 'target programmes', which are special
inter-branch operational plans designed to focus
investment on specific goals and to secure branch
co-operation in such projects (see Cave, 1980,
pp.62-4, 74-6, 85). Among other aspects various
pioneering regions in the north and Siberia,
designated 'territorial production complexes', are
being developed by means of such programmes as
complexes of interconnected economic activity. In
this way it is intended to reduce the dispersal of
investment funds, a process which has always tended
to have negative consequences for transport.

In addition to such long-term planning
policies, various measures have been undertaken to
foster local co-ordination of branch activity. These
measures really date from the post-Stalin period
when there was an attempt to give the republics a
greater role in economic planning and also to
increase the responsibilities of the local soviets.
Further enactments concerning the duties of
republics and local soviets followed in the years
after the 1965 reforms. The essence of these was to
declare that the republics and, to a lesser extent
the local soviets, were to be responsible both for
directly supervising various branches of industry
and economic activity (in the case of the local
soviets, these activities are mainly service and
ancillary functions) and also for exercising
important inspection and control functions. Thus
both republics and local soviets were to watch for
disproportions arising out of proposed new projects
on their territories, such as the disproportions

that might arise from the expansion of an industrial plant on the one hand and shortages of labour, construction industry capacity or transportation capacity on the other. These duties were reinforced by legislation in 1979 and 1981.[9] Naturally, the role of the republics is likely to be much more important than that of the local soviets. The legislation of 1979 and 1981, for example, tightened their control over investment (at least in theory) and ordered them to help in the process of drawing up a new series of territorial production balances.[10] In this way the republics may be in a position to correct or prevent some irrational freight flows. But the legislation also increased the responsibilities of the local soviets, especially in the areas of construction, building materials, and consumer goods. They too, then, may have a role in helping to prevent the unnecessary transportation of certain bulky goods, such as building materials, as well as in other transport matters.

A further facet of the work of the republics and local soviets that requires comment concerns the area of physical planning. Since the 1960s considerable effort has gone into trying to develop and co-ordinate physical planning (see Pallot and Shaw, 1981; Shaw, 1983). Much of this effort has an economic focus and there are important implications for transport. Rationalising passenger and freight transport, for example, is one important goal underlying urban land-use planning, industrial site planning, and settlement system planning. A related activity is the development of industrial nodes and estates with common servicing facilities. Among the economies of scale to be achieved in this process are, of course, considerable economies in transport.[11]

Unfortunately, within the Soviet centrally-planned economy as presently constructed neither long-term planning nor local co-ordination efforts have been found to work particularly well. The problems with long-term planning are well known the world over. Forecasting the future is always a risky business, as difficult in the USSR as it is anywhere else. Even five-year plans have to be revised regularly. Thus the difficulty will be in the actual enforcement of the plans. However, under Soviet conditions, it is probably better to have even a rather unrealistic long-term plan than no plan at all. As far as local co-ordination is concerned, the real problem lies in the relative

lack of power exercised by local soviets and even by
republics in the highly centralised Soviet system.
It is one thing to pass legislation concerning their
rights and duties, it is quite another thing to
ensure that those rights and duties are respected by
others. Thus real co-ordination at the local level
may be an impossible aim in many instances. This is
particularly the case in view of the fact that it
tends to add to the complexity of the planning
process.

Recent legislation has therefore also tried to
promote inter-branch co-ordination through further
centralisation. The July 1979 economic decree, for
example, was certainly a centralising measure.
Several of its provisions were designed to increase
central control over investment and to direct that
investment towards specific goals. These provisions
and also others were meant to enhance both macro-
and micro-balance in the economy and to establish
the five-year rather than the annual plan as the
basic resource allocation document. Stabilisation
of the economy would among other things permit the
transportation ministries to work out optimal
freight-flow schemes for bulk freights based upon
territorial production balances. Needless to say,
many Western economists doubt whether all this can
actually be achieved. (12)

The year 1982 witnessed two other measures
which were designed to foster inter-branch
co-ordination at the regional level, primarily
through central control. At the Party Plenum in May,
Brezhnev announced a series of agricultural measures
to be known as the Food Programme. One key aspect of
this was the setting up at national, republican and
local levels of agro-industrial co-ordinating organs
(forming the so-called agro-industrial complex) to
foster integration of agriculture and associated
industrial processes across ministerial boundaries.
These organs are also to co-ordinate rural and farm
transport in their areas. A programme of investment
in rural roads and infrastructure was announced at
the same time. According to Mozhin, the
agro-industrial integration policy should lead to
the reduction of the often excessive transportation
of foodstuffs through the countryside by road and
rail, the result of the underdevelopment and poor
location of food processing industries (Mozhin,
1983, p.9) (13)

The second measure of 1982 was the approval by
the USSR Council of Ministers in September of a
model statute for representatives of USSR Gosplan in

major economic regions. The representatives are to
chair councils of local planning chiefs and
specialists, and to be supported by a staff of their
own. Among their extensive duties are the fostering
of rational patterns of industrial location, and
supervision over freight flows.[14] That these new
representatives are appointed by Gosplan itself
would seem to suggest that regional planning is now
being taken seriously.

Some Problems of Regional Co-ordination
From what has been said thus far, it is evident that
considerable thought and not a little effort have
gone into the field of regional planning, especially
in the last few years. Although it is still too
early to judge the effectiveness of the latest
measures, perhaps some idea of the nature and scale
of the problems faced in trying to overcome
departmental barriers can be derived from a look at
territorial production complexes. As already
indicated, these are complexes of interrelated
mining, industry and associated economic activity,
primarily being developed in pioneering regions in
the north and in Siberia (see Territorial
'no-proizvodstvennyye, 1981). The solution to their
problems is of considerable importance to the USSR's
economy as a whole, since here lies the future of
the country's fuel and raw material supplies. It is
therefore unfortunate that their development
histories to date still leave a great deal to be
desired.

Foremost in importance among the USSR's
territorial production complexes is the north-west
Siberian oil and gas complex. This region is now
basic to the USSR's energy industry, and its
development is therefore a matter for particular
concern. Ministries central to its overall evolution
are the oil and gas ministries (who frequently fail
to agree between themselves), and there are, in
addition, six other ministries with major
responsibilities who add to the general complexity.
In terms of construction, there is one construction
ministry concerned with oil and gas-related matters,
but numerous other construction agencies and
organisations are also involved. Something of the
flavour of this administrative labyrinth has already
been described with respect to supply systems. In
1980 a special commission of the USSR Council of
Ministers was appointed to supervise the development
of the complex, together with an interbranch

territorial commission of USSR Gosplan, situated in Tyumen'. However, with the 'target programme' and long-term investment plans for the complex still being composed, a lack of co-ordination between the ministries is very evident. This is exacerbated by the fact that Gosplan's territorial commission apparently has limited powers as a planning organ when it comes to the everyday activities of the ministries (Kuramin, et al., 1983). With regard to transport there are, naturally enough, many complaints. For example, the Ministry of Road Transport Construction consistently fails to build roads on time, especially in the newer, more northerly parts of the territory. There was an eight-year delay in the completion of a road to the Medvezhye deposit and facilities had to be constructed in advance of roads, involving considerable extra outlays (CDSP XXXV, no. 31, p.2). There is a current problem with roads at Urengoy, and the prospects further north at Yamburg and Yamal are equally gloomy.

Similar problems are encountered farther to the east along the Baykal-Amur Mainline railway. In this region, a whole series of territorial production complexes has been envisaged following the completion of the railway, and a priority target programme for the development of the zone was ordered in the July 1979 resolution. Again the programme is still being developed and in the meantime the special commission of the USSR Council of Ministers lacks the close supervision needed over day-to-day events (CDSP XXXV, no. 29, p.12). The various ministries, especially the Ministry for Transport Construction and the Ministry of Railways, fail to co-ordinate their activities. The Ministry of Transport Construction is accused, for example, of building a road alongside the railway in unnecessarily temporary fashion and purely for railway construction purposes. After the railway is completed, the road is to be abandoned, but in the opinion of many it is needed to serve the new lumber camps, state farms and also mines (such as the Udokan copper reserves) to be developed in the region (CDSP XXXV no. 35, p.22). Needless to say, because there is no real long-term plan, the ministry has no reason to prolong the life of its road.

Transport is also a problem further north in the new coal mining areas at Neryungri and south Yakutia. This region is regarded as the first territorial production complex in the BAM zone. Here

the Ministry for the Coal Industry tried to
co-ordinate development through a special council
but this failed to exercise the necessary powers
over other agencies. Again, the infrastructure has
lagged behind requirements, and in south Yakutia
there is an over-dependence on shipping along the
River Lena, with associated seasonal navigation
problems and other difficulties (CDSP XXXV no. 2,
p.21).

The position also appears to be unsatisfactory
in other territorial production complexes. In
Krasnoyarsk kray, which contains several nascent
complexes and industrial clusters, the USSR's first
ten-year territorial development plan was drawn up
for 1971-80 (Khorev, 1981, p.186). Yet other sources
reveal that the development of the region is
proceeding without the guidance of a co-ordinating
authority and there are no agreed and integrated
schemes for the TPCs (Mukoyed, 1981). In one or two
places, as along the Yenisey, the planned complexes
have actually dissolved, the separate parts now
being run purely along ministerial lines. The
complex, as one writer put it, thus becomes a
conglomeration (Galushko, 1981). Something similar
seems to have happened on the Kola Peninsula. [15]

Evidently, then, the Soviets still have a great
deal to learn when it comes to the integrated
planning of industrial complexes. Until these
lessons can be learnt, irrational and unnecessary
transportation will continue to be a feature of
these pioneering regions.

The Prospects for Inter-Branch Co-ordination

As the Soviet economy has faltered since the late
1970s, so discussion concerning the prospects and
possibilities of economic reform has become more
widespread. This became particularly noticeable
during the brief Andropov period, when one of the
foci of attention was the perceived necessity of
reforming the ministries in some way and revamping
the whole of the higher industrial bureaucracy.

A typical example of this tendency was an
article by G. Popov which appeared in the journal
Kommunist in December, 1982 (CDSP XXXV no. 23,
pp.1-4). Popov noted the discrepancy between the
structure of ministries on the one hand and actual
branches of the economy on the other. Because of
supply problems, even after 1965, ministries built
up their own ancillary activities and became
'multi-branch complexes', responsible for numerous

types of production. At the same time, as the
economy has evolved, so branches of the economy have
become more interdependent because of the need for
very sophisticated types of technology and for all
kinds of research inputs. Regional problems, such as
natural resource development and the need to
supervise labour inputs, also demand more
co-ordination between branches and between
ministries. Popov pointed out that under the present
system it was difficult for ministries to assume
responsibility for developing those forms of
production which are not central to their concerns.
It is equally difficult for any one ministry to
enforce a uniform scientific and technical policy
for a given branch of production which, in fact also
comes under other ministries. He therefore concluded
by suggesting the need to grant more autonomy and
financial responsibility to industrial associations,
which should replace ministries. The whole
structure, he suggested, should be supervised by a
system of super-ministries responsible for actual
economic branches and with largely administrative
functions. More recently, further articles have
appeared which advocate similar measures but
sometimes with more radical decentralising proposals
as well (e.g. Karagedov, 1983; Kurashvili, 1983).

Some of this discussion naturally has a bearing
upon the problems of transportation. For example,
Popov noted that under the prevailing branch system
it is difficult for any one transportation ministry
to arrange the most efficient combined shipments.
This is one reason why containerisation has not
proceeded more quickly. Some analysts, therefore,
have called for a single all-purpose authority to
co-ordinate the development and work of the
transportation agencies as a whole (e.g. Kolesov,
1982, pp.191-3). At the present time, overall
co-ordination work is in the hands of Gosplan which
has a department of transport and also an Institute
of Complex Transport Problems. But the former is
mainly concerned with integrating the detailed
planning of transport with the national plan,
whereas the latter concerns itself with various
kinds of transport-related theoretical work. As has
already been seen, the various transportation
ministries plan a good deal of their own work
without the help of Gosplan. A single transport
authority would, in the opinion of many, help to
combat departmentalism by supervising the
development and the actual operation of the system
as a whole. It would also have regional sections to

aid co-ordinated development at the local level. At the present time, the various transportation ministries do have their own regional administrations, but often these correspond neither with each other, nor with the boundaries of single republics or oblasts.

Conclusion

Departmentalism in Soviet transportation, and in the economic structure as a whole, results from the country's current ministerial system. At present the central authorities are attempting to combat it by giving more power to the centre and by trying to make planning more efficient, both nationally and regionally. There is the prospect of attacking the problem by more fundamental reform, involving a shake up of the ministerial structure. This could involve, for example the founding of super-ministries, or the granting of more power to regional authorities who could then become responsible for planning industry and transport within their areas.

It is, of course, the opinion of most Western economists that such measures cannot really achieve much. The basic faults in the system, in their view, are not such things as departmentalism. If this alone were abolished it would be replaced by something else equally unhelpful, such as localism. The real faults are overcentralisation and lack of incentive and competition. Not until these things are put right can much progress be hoped for, and there is little prospect at the moment for such a radical change in course.

On the positive side, however, the continuing experimentation and adjustment in the system do at least show a willingness to face up to some of the realities which result from a complex, industrial economy. Whatever may be the failings in transportation when it comes to the drive for efficiency, the Soviets are likely to learn some lessons from their present strivings in the problem-ridden areas of transport and regional co-ordination. These are, after all, problems which beset all industrial economies, and not one of them has yet found a comprehensive solution.

NOTES

(1) FBIS, 'Proceedings of the 26th CPSU Congress', volume I, p.25. Similar sentiments were

expressed by Yuriy Andropov at the Party Plenum in November, 1982, although the situation seems to have improved a little during 1983 and 1984.

(2) The figures are for all forms of transport, apart from roads. Mozhin (1983, p.7) reports that the annual cost of transporting fuel for 3-4,000 kilometres from Siberia to the European part of the country now exceeds 2,000 million rubles.

(3) See Lavrishchev, 1968; Nikol'skiy, 1979, p.5, and references quoted in Mellor, 1975, pp.75-6.

(4) Werner Gumpel, quoted in Mieczkowski, 1978, p.45.

(5) See Ekonomicheskaya gazeta, 1979, no. 32, article 28.

(6) Kolesov (1982, p.178) also noted that each year the West Siberian railway has to turn away between 4 and 7 million tons of building materials because it is unable to ship them.

(7) Kolesov (1982, p.198) claims that losses to the Soviet economy resulting from transport inefficiencies now exceed investments in transport by approximately 8-10 times. This, however, seems excessive.

(8) For references to some Western work on this problem, see Pallot and Shaw, 1981, p.138.

(9) For the 1979 resolution, see Ekonomicheskaya gazeta 1979 no. 32, pp.9-17. For the 1981 resolution on furthering the role of the soviets in economic construction, see Sobraniye postanovleniy pravitel'stva SSR, otdel pervyy, 1981, no. 13, st. 78, pp.350-4.

(10) Mieczowski (1978, p.149) suggests that there has up to now been no central planning of interrepublican freight flows - only all-Union ministries have planned their own interregional flows.

(11) See Ekonomicheskaya gazeta 1984 no. 28, p.10.

(12) The 1979 resolution is examined in Schroeder, 1982, and Hanson, 1983. Among transport-related provisions not already mentioned, the resolution instructed Gossnab to revamp its transport schedules according to an agreed and centralised system. Further, it called for a new scheme of material penalties to be levied on transportation agencies for the non-fulfilment of freight-delivery contracts.

(13) Mozhin also mentions the excessive size of food processing plants (convenient from the view-point of the administering ministries but not from

that of agriculture) and the lack of agricultural
specialization within regions as factors requiring
attention.
(14) See Sobraniye postanovleniy
pravitel'stva SSR, otdel pervyy, 1982 no. 28, st.
145, pp.531-6.
(15) Article by N. Nekrasov in Pravda,
October 19th, 1982.

REFERENCES

Biryukov, V. (1983) 'Transportnaya sistema strany v
     sovremennykh usloviyakh', Planovoye
     khozyaystvo, no. 12, 3-10
Cave M. (1980) Computers and Economic Planning : The
     Soviet Experience, Cambridge University Press,
     Cambridge
CDSP : Current Digest of the Soviet Press, Joint
     Committee on Slavic Studies, Washington DC
Dmitrieyev, V.I. (1977) Teoreticheskiye problemy
     postroyeniya gruzovykh tarifov transportnoy
     sistemy SSSR, 'Transport', Moscow
Galushko, Yu. I (1981) 'Kak raspalsya kompleks',
     Eko, no. 3, 88-91
Gustafson, Thane (1981) Reform in Soviet Politics :
     Lessons of Recent Policies on Land and Water,
     Cambridge University Press, Cambridge
Hanson, Philip (1983) 'Success Indicators Revisited;
     The July 1979 Soviet Decree on Planning and
     Management', Soviet Studies, 35, January, 1-13
Karagedov, R.G. (1983) 'Ob organizatsionnoy
     strukture upravleniya promyshlennost'yu', Eko,
     no. 8, 50-60.
Khorev, B.S. (1981) Territorial'naya organizatsiya
     obshchestva, 'Mysl'', Moscow
Kolesov, L.I. (1982) Mezhotraslevyye problemy
     razvitiya transportnoy sistemy Sibiri i
     Dal'nego Vostoka, 'Nauka', Novosibirsk
Kuramin, V.P. et al. (1983) 'Problemy planirovaniya
     i upravleniya razvitiyem Zapadno-Sibirskogo
     neftegazovogo kompleksa', in Problemy razvitiya
     Zapadno-Sibirskogo neftegazovogo kompleksa,
     'Nauka', Novosibirsk, pp.531-6
Kurashvili, B.P. (1983) 'Sud'by otraslevogo
     upravleniya', Eko, no. 10, 34-57
Lavrishchev, A. (1968) Economic Geography of the
     USSR, Progress, Moscow
Lebedeva, L.P. (1983) 'Sistema
     material'no-tekhnicheskogo snabzheniya
     Zapadno-Sibirskogo neftegazovogo kompleksa' in
     Problemy razvitiya Zapadno-Sibirskogo

neftegazovogo kompleksa, 'Nauka', Novosibirsk, pp.185-199

Materialy XXVI S'yezda KPSS (1981), 'Politizdat', Moscow

Mellor, R.E.H. (1975) 'The Soviet Concept of a Unified Transport System and the Contemporary Role of the Railways' in Leslie Symons and Colin White (eds) Russian Transport, Bell, London, 75-105

Mieczowski, B. (1978) Transportation in Eastern Europe, East European Quarterly, Boulder

Mozhin, V. (1983) 'Ratsional'noye razmeshcheniye proizyoditel'nykh sil i sovershenstvovaniye territorial'nykh proportsiy', Planovoye khozyaystvo, no. 4, 3-12

Mukoyed, A.L. (1981) 'Rol' mestnykh Sovetov v razvitii territorial'no-proizvodstvennykh kompleksov', Sovetskoye gosudarstvo i pravo, no. 2, 12-19

Nekrasov, N.N. (1975) Regional'naya ekonomika, 'Ekonomika', Moscow

Nikol'skiy, I.V. (1978) Geografiya transporta SSSR, MGU, Moscow

North, Robert (1979) Transport in Western Siberia, University of British Columbia Press, Vancouver

Nove, Alec (1980) The Soviet Economic System, Allen & Unwin, London

Pallot, Judith and Shaw, Denis J.B. (1981) Planning in the Soviet Union, Croom Helm, London

Pavlenko, V.F. (1975) Territorial'noye planirovaniye v SSSR, 'Ekonomika', Moscow

'Problemy uluchsheniya kachestva planirovki i zastroyki novykh gorodov' (1978), Arkhitektura SSSR, no. 2, 28-45

Schroeder, Gertrude (1982) 'Soviet economic "reform" decree : more steps on the treadmill', in Soviet Economy in the 1980s : Problems and Prospects, U.S. Congress, Joint Economic Committee, Washington DC, Part 1, 65-88

Shafirkin, B.I. (1975) Transportnyye zatraty narodnogo khozyaystva i puti ikh sokrashcheniy, 'Transport', Moscow

Shaw, Denis J.B. (1983) 'The Soviet urban general plan and recent advances in Soviet urban planning', Urban Studies, 20, 393-403

Smolyar, I. (1976) 'Realizatsiya general'nykh planov kak osnova upravleniya razvitiyem goroda', Arkhitektura SSSR, no. 5, 13-17

Territorial'no-proizvodstvennyye kompleksy SSSR (1981), 'Ekonomika', Moscow

Chapter 2

SOVIET RAILWAYS - LETHARGY OR CRISIS?

John Ambler, Holland Hunter, and John Westwood

What is wrong with the Soviet railway system? With some effrontery, this contribution, an amalgam of studies made over recent years, attempts to decipher the messages from the MPS (the Ministry of Railways) and Gosplan, and hence to answer this question. Since the railway, known internationally but not domestically as SZD, moves about 55% of the world's rail freight and 25% of the world's rail passenger traffic, the question is not an easy one.

Hunter's research over 30 years has had as its main theme 'how much rolling stock - freight cars and locomotives - does the system possess, and is it adequate for the traffic?' Westwood's specialism is engineering, with particular emphasis on technical advances in rail operations. Innovations on the SZD have to be set in the context of the bureaucratic structure of the USSR and the Russian psyche, both somewhat alien to westerners. Ambler's interests lie in the commodity flows and geographical dispersion of the rail system, together with its management and finance. It will be clear that productivity is the area where our three perspectives coincide.

The railway problem is simply stated : it is that freight traffic between 1975 and 1982 grew at only 1.0% per annum, against 7.0% per annum in the quarter century between 1950 and 1975. Many key productivity indicators declined - cause or effect? - and the abundant literature became ever more introspective. Five indicators are cited in Table 2.1 to show the seriousness of the SZD's position.

Over the last 10 years, the supply of hard statistics on overall operations, as distinct from vignettes for purposes of propaganda and exhortation, has gradually dried up; in any case, the figures available were for the most part 'derived statistics' (we know average train weights

Table 2.1 Annual average rates of change (AARC) of key indicators

	1970-75	1975-82
Freight train speed	+1.3%	-0.9%
Freight train weight (gross)	+2.6%	+0.6%
Daily kilometres per freight car	+2.1%	-1.5%
Profit (labour rate corrected)	+7.0%	-8.6%

but not gross ton-kilometres and train-kilometres from which that statistic is derived). Particularly useful are the railwaymen's newspaper Gudok (The Whistle) and the technical monthly Zheleznodorozhnyy transport : when these are quoted we give year, month and day or page. The main books used as sources are listed in the bibliography; they are quoted so extensively that we do not give every page reference.

The answers to the problems facing the railway are clear and simple to the Soviet leadership but not to the railway managers, who do not view with equanimity their new roles, not unlike Boxer's in Animal Farm. Some critical decisions will have to be made, and we examine their background. Who will win in any conflict between Gosplan, the MPS, and the other ministries? Will there be more railways as a sop to decentralisation and local minorities or fewer, with advantages from economies of scale? Is it better to have a myriad of small improvements (better availability of spares, for example) rather than a mammoth project such as the BAM? Does the railway suffer from congestion, and if it does, where? Is staff morale low, and if so, what might raise it? Does a railway have an inexhaustible supply of inner reserves which can be tapped if only management applies itself? Should there be one, two, or dozens of indicators of railway productivity? Finally, is the railway failing the economy, or is the economy itself faltering?

Some notes by way of preface will hopefully guide the reader through this maze. First, the railway terminology we use is largely but not exclusively American, if only because their usage is more lucid to the lay reader. Thus we have 'switches' rather than the ambiguous 'points' and 'freight car' rather than 'wagon', but we use 'open car' instead of either the American 'gondola' or the

Soviet Railways

English 'mineral wagon'. We give the word 'traffic'
a technical sense, namely ton-kilometres (tkm) or
passenger-kilometres (pkm) according to context; and
the concept of 'annual average rates of change' is
used so often that it is abbreviated to AARC. Other
abbreviations and terms are explained in the
glossary. Second, the lengthier tables are relegated
to the end of the chapter, the text itself
containing only summarised data in tabular form.
Annual figures before 1965 are available from
several English sources, notably Williams (1962) and
Hunter (1968). 1969, 1970, and 1975 are quoted more
frequently simply because a larger volume of
consistent data is available.

The metric system is used exclusively,
'billion' means 'thousand million' (milliard),
'diesel' means more precisely 'diesel-electric', and
'railway' denotes either the SZD as a whole, or one
of its constituent parts, now 32 in number; on
average each of these is about the same size in terms
of freight traffic as the French and German systems
combined, or five times the British system.

Wider Aspects of the Problem

Importance of Rail in the Soviet Economy. The
distribution in 1980 and growth over 25 years of
internal freight traffic is shown in Table 2.2.

Rail's share dropped from 86% to 64%, still far
in excess of the shares for North America and
Western Europe. In 1955 the USSR accounted for 37%

Table 2.2 Distribution and growth rates of freight
traffic by mode within the USSR

	billion tkm (1980)	AARC 1955-80
Rail	3515	+5.2%
Road	430	+9.7%
Inland waterway	250	+5.3%
Cabotage	70	+2.4%
Pipeline	1215	+19.3%
	5480	+6.4%

of world rail freight traffic; by 1980 this had grown to 52%. At any one time, 40 million tons of various commodities are somewhere on the rail system.(Total domestic transport demand per year per inhabitant is about 22,000 tkm, compared with 17,000 in the USA, and 3,000 in West Germany.)

The growth in rail traffic in the Soviet Union since the revolution has been substantial and well nigh continuous. Compared with the pre-revolution peak in 1916, the railway now transports forty times as much freight. Only six times has a year, after correcting for leap years, shown a decline in traffic, twice in wartime, and in 1920 and 1933. The shocks of 1979 and 1982 must therefore have been severe : the drop in 1979 was about the same as the entire year's rail freight of West Germany and Italy combined. The importance of the railway to the economy can be assessed (Table 2.3) from the average shares by commodity of freight traffic (1980-2) and 50 year (1932-82) AARCs :

Table 2.3 Commodity shares and growth rates (rail traffic)

	Share	AARC 1932-82
MBM	13.5%	+7.6%
Ores	6.9%	+7.4%
Iron/Steel	8.1%	+7.3%
Oil	13.2%	+7.0%
Other	30.0%	+6.4%
Coal	17.0%	+5.7%
Timber	7.3%	+4.6%
Grain	4.0%	+4.3%
(MBM = mineral building materials)		

90% of freight is heavy bulk commodities essential to any 'smoke stack' economy. ALH (average length of haul) is very long by European standards, 930 kilometres being the straight line distance between Berlin and Paris.

The Winter of 1979. Over the years, the share of traffic in the first three calendar months has gradually approached the 24.7% represented by their 90 days. In 1950 it was 21.0%, in 1960 23.7%; between 1975 and 1982 it never fell below 24.2% except in 1979, when it was a disastrous 23.4%. Climatic conditions in the USSR are much worse than

in Europe and the USA, though Canada's are nearly as
bad. Very approximately, in the average January, 10%
of traffic moves in areas below - 20°C, and 55% in
areas between - 10° and -20°. The mean temperature,
rail freight traffic weighted, is -13°.

In the first three months of 1978, average
daily tkm was 9.3 billion, dropping to 8.7 billion
in the corresponding period of 1979. The loss of
traffic in those months was 52 billion tkm, and 28
billion in the rest of the year. Table 2.4
summarises about 25 weather station records for 1979:

Table 2.4 Weather variations in 1979 from average
conditions.

	Temperature	Precipitation
January	-2.2°C	+20mm
February	+1.6°C	+13mm
March	+3.2°C	+ 3mm

In general, the more southerly areas experienced
deeper snow. Moscow had 73 mm in January against an
average of 36 mm : it snowed on fourteen days.
Rostov-na-Danu had 2 metres in the three months,
over 32 days. Other cities with over double the
average January snowfall included Kazan', Voronezh,
L'vov, Perm', Sverdlovsk, Orsk, Krasnoyarsk,
Barnaul, and Blagoveshchensk.

The analysis by republic and type of freight is
significant. The RSFSR lost 64 billion tkm compared
with 1978, and the Ukraine 9 billion. In terms of
shipments, the RSFSR lost 13.8 million tons of oil
(the USSR as a whole lost 4.4 million), and 19.8
million tons of timber, plus a staggering 25.3
million tons of MBM (USSR - 39.0 million), about
4.9% of its loadings of that commodity in 1978.

Since the greatest decline came in the RSFSR,
production figures for that republic were analysed
by area. On an unweighted basis, the biggest
relative declines starting with the largest were in
Kuybyshev, Tambov, Mariy, Volgograd, Ivanovo,
Kemerovo, Chita, Sverdlovsk, Tyumen', Penza, Moscow,
Altay, Saratov, Leningrad, and Maritime ASSRs, krays
and oblasts. The correspondence with the cities with
particularly severe weather noted above is close.
ZhT (80/2/56-7) confirms this : 'the breakdown of
train scheduling in the low temperature period of
end '78 and beginning '79 demonstrated weak links. The
problems that appeared everywhere were caused not

only by individual equipment failures but because of
the inadequacy of yards, whose tracks became clogged
with rolling stock, preventing snow clearance etc.'

Profitability of the Railway System. In 1974, the
first year when total profits in the USSR exceeded R
100 billion for the first time, the profit for all
rail enterprises was 6.44 billion; rail traffic
produced 6.15 billion of this. Profit on capital was
12.9%. (In this section, profit is gross, i.e. after
amortisation but before deducting interest even on
bank borrowing.) Table 2.5 details various
statistics, where appropriate adjusted crudely for
price changes. Profit (A) is expressed in thousands
of personnel, calculated at economy-wide pay rates
plus social charges. (In 1970-75, rail profits were
6.4% of national profits, in 1982 only 3.4%.)
Average costs of freight (B) and passenger traffic
(C) are similarly expressed, the denominator being
in billion tkm and pkm. Investment as a share of
profit is shown in (D). (E) gives the excess of rail
pay over average pay; (F) gives total labour
productivity of the SZD, measured in thousand tkm
and pkm combined per employee.

Table 2.5 Railway finances and labour productivity.

	(A) Profit	(B) Freight Cost	(C) Pass. Cost	(D) Invest-ment/ Profit	(E) Rail premium	(F) Output per Worker
1960	2308	2136	4437	45.2%	2.9%	833
1965	2847	1549	3586	36.5%	2.3%	1087
1970	2919	1186	2763	40.2%	1.1%	1381
1975	2888	1038	2540	44.2%	8.4%	1714
1976	2801			44.3%	5.6%	1735
1977	2631			47.6%	8.5%	1748
1978	2512			53.4%	7.5%	1768
1979	2190			62%	6.8%	1719
1980	1982	1015		70%	11.1%	1722
1981	1906			75%	11.1%	1735
1982	1543			90%	11.7%	1709

The figures after 1975 make it clear that even
though more workers have been employed, enjoying an

ever-increasing pay differential, output per man is
on a plateau, profits in total buy less and less
(after an all time peak in 1967), and a larger and
larger share of it is required for the railway's own
investment: in fact, the lion's share of investment
funds comes from amortisation. Worthy of note is the
fact that SZD's operating ratio between 1965 and
1975 was much the same as on railways in Britain
between 1875 and 1895.

Table 2.A gives data on individual railway
profits and other key figures. The profit decline
between 1969 and 1975 applied to all 26; a
significant problem is that the bulk of profit comes
from those railways which, carrying as they do the
heaviest traffic, find it hardest to carry any more.
In 1975, the West Siberian and the two Urals
railways, which had 11.7% of the route, produced
21.3% of the profit; their counterparts in European
Russia (Gor'kiy and Kuybyshev) had 7.3% of the route
and 12.6% of the profit.

It is easy to record costs for a particular
railway making up an integrated network, since in
principle only locomotive hauled rolling stock moves
across railway borders, and even then it is only its
amortisation (about 6-7% of the total cost) which is
not specific to an area. The allocation of revenue
is much more complicated. 57% of passenger traffic
in 1979 was shared between two or more railways, and
84% of freight traffic in 1975. The distribution
between railways is effected by the MPS and the
method alters in minor respects from time to time.
Broadly, the movement-charge revenue is pro-rated by
distance moved, and for the terminal charge a flat
rate per ton is allowed. Supplementary revenues for
weighing, demurrage, storage, and various terminal
services lie with the railway where the service is
performed : in 1973, Silayev (1975, pp.37-8) shows
that R 2091 million was earned by local freight
traffic, 8614 million in cross railway movement, 965
million in shared terminal charges, and 336 million
from supplementary revenues.

Although freight traffic is still highly
profitable by Western standards, if much less than
it used to be, passenger traffic is less so - 32% of
revenue in 1975 against 39% on freight. The ratio
between freight and passenger cost per tkm and pkm
has been rising steadily, from 1.63 in 1950 to 2.45
in 1975. Kovrigin and Belen'kiy together show that
suburban traffic, an ever-increasing element of the
total, was in fact unable to cover its costs in 1975
by 2%, because season ticket traffic (55%) produces

revenue which only covers a quarter of its total
cost. (On the six metros, expenses in 1980 absorbed
98.6% of all revenue, an achievement which the MPS
must wish could be matched for rail suburban
traffic.)

Productivity of rail assets

Rolling Stock Production and the Working Park. In
volume terms, output in the USSR in the post-war
period has been vast. By the middle '60s, it had for
example a quarter of the world's locomotive
production. Production figures for 5-year blocks are
given in Table 2.6.

Table 2.6   USSR rolling stock production since 1945

	Electric locos		Diesel sections		Steam locos	Freight cars
	No.	HP(000)	No.	HP(000)	No.	(000)
1945-9	137	442	222	223	3144	116.0
1950-4	630	2059	497	914	3330	152.6
1955-9	1459	5939	2409	4816	1144	191.8
1960-4	2851	15933	7244	14860	-	183.8
1965-9	2230	14995	7475	17655	-	221.9
1970-4	1727	12899	7307	19290	-	335.1
1975-9	2079	16170	6901	19172	-	345.9
1980-2	1290	10584	4033	11186	-	182.6
Total	12403	79021	36088	88116	7618	1729.7

(Two diesel sections are equivalent to a main line
'locomotive')
     In this same period, about 62,000 passenger
cars were built, and quantities of electric
locomotives for passenger service imported from
Czechoslovakia. The number of switchers produced was
somewhat below 15,000. For passenger traffic in
suburban service multiple units, mostly electric,
are available. Between 1965 and 1979 alone, 109,000
freight cars were imported, in addition to the
figures in Table 2.6. (These figures are thought to
exclude those for industrial use.)
     Communist economics have for long reported
railway statistics using the 'working park' as the
denominator. The non-working park (reserve) covers

vehicles in need of major repair, plus those used for secondary purposes and lying in strategic reserves. Table 2.B contains data on the freight car (and locomotive) stock, culled from a wide variety of sources. No matter which way one manipulates the figures, the result is a working park of between 1.2 and 1.3 million freight cars; there are grounds for thinking that the proportion in reserve has grown over the last decade, so that the total stock is about 1.9 million, with a replacement cost of R 15 billion.

The locomotive working park is another matter, however. A good starting point is Gundobin (1971, p.513), whose percentages are used to derive Table 2.7.

Table 2.7 Locomotive working park in 1970 by type of service.

	Electric	Diesel	Steam
Freight	4300	5200	100
Passenger	1350	1550	50
Service	200	400	50
Switching	150	3700	250

(In 1975, there were also about 1800 electric and 250 diesel multiple unit sets in the working park.)

The real problem for us 'outside' analysts arises with diesels. From the production and import figures to 1982, one would expect at least 30,000 sections to be in stock, and hence about 25,000 in the working park. Calculating back from diesel's share of traffic, and train weights, produces a figure of only 14,000, assuming two sections per train; this leaves an unbelievable 8,000 twin sections in reserve. (Table 2.7 is also affected by this confusion between sections and locomotives.) For every four MPS freight cars and two MPS locomotives, industrial enterprises have one : how many of these are in the production figures?

However, our present interest is not so much the actual numbers in stock, but whether there have been significant changes in rolling stock productivity over the years. Provided there are no changes of definition, we can observe such changes from the various derived statistics even though we cannot examine the basic data (numbers of cars, train-kilometres and hours, etc.)

First (technological) Revolution. Soviet railways have been able to save themselves from disaster twice before, the first time being in the early '30s. To describe what happened in the past, and what Gosplan is hoping will happen again, it is necessary to explain the mathematics underlying freight performance. Two identities exist, relating to freight cars and locomotives separately : the first is, traffic = number of cars x average capacity/car x capacity utilised x loaded running percentage x movement speed x hours of movement/car; the second is, traffic = number of locomotives x train weight x train speed x movement hours/locomotive. If traffic can be increased without increasing the physical numbers either of cars or locomotives, there has been a mobilisation of what are termed internal or hidden reserves; for brevity we use 'inner reserves'.

Substantial spare capacity has always existed on Western railways, and has been heavily drawn on in wartime. The AARC in traffic in Britain between 1938 and 1944, the all-time peak for rail freight, was +7.0%, and this was achieved with an increase in stock of only +0.2%.

The first revolution on the SZD took place between 1929 and 1937, and is unjustly associated with Kaganovich. Most references are merely to the fact that traffic in 1933 was the same as in 1932. Table 2.8 gives the freight car contribution and Table 2.9 the locomotive contribution to traffic changes over the whole period as well as 1932-3; British wartime comparative figures are also given.

Table 2.8 Freight car productivity factors and traffic increases pre-war (AARC)

	USSR 1932-33	USSR 1929-37	Britain 1938-44
Traffic	+0.4%	+15.4%	+7.0%
Freight car numbers	+1.9%	+6.0%	+0.2%
Capacity (each)	+1.3%	+1.1%	+0.6%
Capacity utilised	-1.9%	+1.2%	+1.2%
Loaded car-km percentage	-1.2%	+0.3%	+1.7%
Speed in trains	-3.1%	+5.1%	-4.1%
Movement hours/car	+3.5%	+0.9%	+7.5%

Soviet Railways

Table 2.9 Locomotive productivity factors pre-war (AARC)

	USSR 1932-33	USSR 1929-37	Britain 1938-44
Locomotive numbers	+2.8%	+3.5%	+2.7%
Train weight	-1.8%	+5.2%	+4.4%
Train speed	-3.1%	+5.1%	-4.1%
Movement hours/locomotive	+2.4%	+0.9%	+4.0%

Over the eight year period, faster and longer trains were the prime contributors to carrying the extra traffic, but without the additional stock of both cars and locomotives the economy would have been held back. Likewise, the deterioration in 1933 can be ascribed to the same two factors. In 1929-32, rail investment accounted for 10.2% of the USSR total, and in 1933-7) its share rose to 10.7%; the average from 1956 to the present time has been just under 3%. The most notable single investment seeding the first revolution was the class FD 2-10-2 locomotive, with its heavier axle load enabling it to contribute so much to raise train weights and speeds.

The Second Revolution. The second revolution was based on what the literature calls 'progressive traction types', i.e. the switch from steam to electric and diesel haulage. Steam carried on a long time, and still accounted for 0.1% of freight

Table 2.10 Traffic and haulage type

	1958	1965	1970	1975	1982
Tariff tkm (billion)	1302.0	1950.2	2494.7	3236.5	3464.5
Reported tkm (billion)	1319.4	1982.2	2561.8	3307.2	3542.5
Electric	15.1%	39.5%	48.7%	51.7%	57.8%
Diesel	11.3%	45.0%	47.8%	47.9%	42.2%
Steam	73.6%	15.5%	3.5%	0.4%	0.0%

traffic and 1000 km of route in 1979, but the main benefits of the changeover took place in a few years each side of 1960. The availability of detailed figures for 1958 and 1965 permits a quantification of the benefits : Table 2.10 shows traffic and method of haulage, and Table 2.11 the contribution made by locomotives.

Table 2.11 Locomotive productivity factors 1958-1965 (AARC)

	Total	Switch from Steam	Straight Increase
Traffic	+6.0%		
Locomotive numbers	-3.3%		
Train weight	+2.4%	+1.1%	+1.3%
Train speed	+2.4%	+0.6%	+1.7%
Movement hours/locomotive	+4.5%	+3.7%	+0.8%
Output per locomotive	+9.6%	+5.5%	+3.9%

The 'straight increase' column records the productivity benefits which are due to more performance from a static mix of locomotives. Thus, electric train average speed increased from 43.7 km/hour to 50.1, and net weight from 1300 to 1455 tons. Even steam-hauled trains moved faster, though they became lighter. Output per locomotive is calculated in the standard way, i.e. net tkm per

Table 2.12 Freight car productivity factors 1958-1965 (AARC)

Car numbers	+4.1%
Capacity (each)	+1.7%
Capacity utilised	-2.1%
Loaded car-km percentage	+0.1%
Speed in trains	+2.4%
Movement hours/car	-0.3%
Output per freight car	+1.8%

(Freight car output is calculated as net tkm per freight car day in Table 2.12.)

locomotive day. The AARC in this basic statistic was
electric +4.8%, diesel +5.0%, steam -0.7%. Increased
speed of movement, it must be noted, is the only
factor common to both locomotive and freight car
productivity (Table 2.12).

During the second revolution, steam locomotives
were available for train haulage about 4 hours a day
less than electrics and diesels (Izosimov 1967,
p.49); this, coupled with the other factors of
weight and speed, caused steam locomotive
performance in 1965 (see Table 2.B) to be 224,000
(electrics : 727,000, diesels : 654,000). Rolling
stock accounted for 44% of total investment during
the second revolution. This time, however, car
numbers were the leading contributor to productivity
increases, setting aside the greater hours of work
of 'progressive' traction locomotives. Between 1958
and 1965 (inclusive), 302 thousand were built, at a
cost of around R 1.2 billion; the main line freight
locomotive investment was some R 2.2 billion,
although, of course, the decline of steam produced a
drop in total numbers of locomotives.

Freight Car Utilisation. Various sources (ZhT,
Dmitriyev 1980) show clearly how the pressure on
freight car usage has increased. The percentage of
capacity used (static load) increased from 1965 to
1975, starting at 79.2% and ending at 85.1%. Each
increase of x% is of course marginally more
difficult than the previous such increase. Table 2.B
gives some of the principal data available directly
or indirectly. After 1965, all-time productivity
peaks occurred in the following years : 1974 - net
daily output i.e. tkm per freight car and per
capacity ton, 1971 - daily hours in movement, 1971 -
daily km per freight car, 1965 - turnround time in
days, 1975 - loaded running percentage.

One problem is that specialised freight cars,
by definition, have less availability for traffic.
The fleet is divided into 6 parts, the initial
number denoting the type : thus tank cars are
numbered 7xxxxxx, box cars 2xxxxxx. In 1970, no less
than 42.5% of the total fleet (and 42.5% of the
working park) consisted of open cars, and only 4.4%
(3.8%) of special cars. Sergeyev (1975, p.63) shows
just how universal is the open car's use : 90% of
iron and steel, three quarters of timber, half of
brick shipments, and a fifth of fertiliser traffic
moved in them. They are often not suitable for the
freight offered, since it is exposed to weather and

theft, but this practice helps to ensure that output per open car is over a quarter greater than the all-fleet average; in the early '70s open cars carried 52% of all freight traffic, and in 1975 they were loaded on average to 91.6% of capacity of around 64 tons per 4-axle car.

Box, refrigerator, and flat cars are at the other end of this scale. In the 1970s they accounted for 38½% of the working park, 33% of car kilometres, but only 28% of traffic. In 1975 static loads were 60-65% of capacity. Output per car was three quarters of the all-fleet average.

The specialised freight car, mostly owned in the West by rail customers, has an intermediate position. Such cars, numbered in the series 9xxxxxx, form a mixed group; cement cars cannot be used to move containers, trucks, or grain, and vice versa. The price for greater suitability to the traffic is a reduced loaded running percentage, though speed of loading and movement may be higher. Sergeyev (1975, p.98) gives percentages for that statistic : cement - 55.6%, grain - 58.8%, containers - 55.6%; the worst all-fleet statistic since the war was 70.8% in 1958, the best 73.1% in 1945 (there are no figures after 1975). Specialised cars were loaded in 1975 to 94% of capacity, on average. As a result, daily output of a cement car was 3%, and of a grain car 24%, above average, but other types brought the overall ouptut of all specials more than 10% below the all-fleet average.

There is one factor tending unobtrusively to improve freight car productivity. Between 1950 and 1982, ALH increased from 721.9 to 930.1 km, at an annual average increase of 0.8%; in only eight years was it lower than in the previous year. This fact alone improves freight car productivity, since the proportion of a car's time moving in trains is increased. During the early 1970s, the annual increase of ALH of 0.8% resulted in an increase in output per freight car of around 0.55%, thus postponing each year the need for over 5,000 freight cars. These calculations take into account the increase in gamma factor, and allow for a modest increase in time at marshalling yards.

A counter factor was also at work, however. Povorozhenko (1974) reveals long term growth in LCL traffic, and articles in ZhT the same in container shipments. LCL traffic is the bane of Western railways, where it costs far more than it can earn because of its intensive labour needs and low loadability. In Switzerland, revenue covers less than 40% of cost, against 75% for car load traffic, while accounting for one in five loaded car-kilometres; the dynamic axle load is about three quarters of a ton. In the USSR, ALH is about double the national average, and dynamic loading about 3½ tons per axle; nevertheless, the strain on productivity is severe. In 1950, LCL was 0.32% of total tonnage, in 1970 0.41%. In that year, one in every 40 loaded car-kilometres was carrying LCL.

Container traffic is similar to LCL traffic. In the mid '70s, four out of five containers were of 3 tons capacity, and were frequently loaded on open cars, which will take twelve with careful placing. Such a car, at average container loading, will have a static load of just under 30 tons. Container traffic amounted to 0.34% in 1950, 1.32% in 1960, and 2.05% in 1975, by which time large containers were taking 12% of it.

Both these types of traffic are afflicted by trans-shipment, about twice in the case of the average movement, which extends transit time and holds up freight cars at the trans-shipment points. This apart, the effect of the relative increase in LCL and container traffic is to reduce freight car traffic output each year by about 5 tkm a day, significant if output is for other reasons at a maximum.

Freight Train Operations. The running of freight trains on the SZD is much the same as on other railways, scale apart. Typically, for a long distance movement, a loaded car is collected at a loading point by a pick up freight train and taken to a local yard, where it is sorted and taken away by a local freight for re-sorting at a main yard; it moves between main yards at high speeds in lengthy freight trains; at the receiving end it goes through the reverse cycle, but after unloading, over 80% of cars have to be collected empty and distributed to a new loading point. There is a trade-off in yard operations : if there are too frequent train movements between yards, transits will be faster but trains shorter. There is a limit to the number of yards which can receive trains from any one yard, set by the number of individual sorting and departure tracks; hence, cars which might otherwise move in separate trains to separate yards may require re-sorting down line. Yard practice is further complicated by the fact that main yards also have to deal with some local traffic.

38

This applies to car-load traffic only. Marchroute traffic (a very approximate English equivalent is unit or block train loads) avoids some or all of such marshalling cost and delay, as does traffic in express freights, which travels particularly long distances : Soviet LCL and refrigerated traffic has an ALH of nearly 2000 km.

Abramov (1974, pp.128-30) gives insights into these activities, and Table 2.13 reworks the data for 1975. Some of the figures given there are frankly speculative; in particular, the figures for marchroute traffic use the broadest definition. In addition to the locomotive numbers estimated there, which include assisting and light running as 'other', another 2000 or so were allocated to work at way stations, collecting and depositing freight cars. 15% of electric, 20% of diesel, and 35% of steam locomotives were on such duties. Abramov (1974, p.117) indicates that 28% of switcher hours in 1971 were in yards, 36% at way stations, and 36% on industrial sidings.

Table 2.13   Freight train operations in 1975

	Annual car km (billion)	Train km (billion)	Daily loco km	Locomotives (moving and at depots)	
				Principal train locos	Other train locos
Express	3.5	0.1	610	420	30
Marchroute	35.0	0.7	600	2870	330
Through	30.0	0.45	540	2020	260
Local	23.0	0.45	520	2140	230
Transfer	2.5	0.05	330	390	30
Trip	1.5	0.05	290	440	30
Pick up	5.0	0.15	320	1180	100
	100.5	1.95	510	9460	1010

Abramov (1974, p.133) also gives data on the average leg between car operations by train type which forms part of table 2.14.

The benefit from marchroutisation, which accounted for 34.8% of tonnage in 1965 and 46.5% in 1976, can also be seen from Abramov. Of total freight cost, 13.9% was related to technical

stations, where 35% of time was spent. The average
leg between yards was 318 km in 1958, and about 355
km in the late seventies.

Table 2.14   Freight   car   marshallings   en   route
(1972)

	Leg (km)			Average trip	
	Marshalling	Transit	Yard	Times marshalled	Car km
All freight trains	352	199	127	4.03	1417
All through trains	530	224	157	2.43	1288
Transfer and trip trains	56	42	24	1.00	55
Pick up trains	127	-	127	0.60	74

In Britain, the emphasis on running freight in
train-load quantities has lifted the share of
tonnage moved thereby from 33% to 91% in 20 years;
over the EUROP countries the proportion is 50-55%.
However, the loss of traffic has been severe, and

Table 2.15 Marchroute transit time advantage

	Days saved as against all tonnage		Marchroute share				
	500-999 km	1000-1599 km	All	Coal	Oil	Ores	MBM
1950	1.28	1.85					
1955	1.13	1.42					
1965	0.54	1.20	34.8%				
1970	0.73	1.45	39.2%				
1975	1.06	1.35	46.1%	62.3%	81.3%	86.5%	49.2%
1979	1.3	1.3					

(1979 figures from Kozlov & Polikarpov [1981])

too late it has been realised that the high fixed costs of the route have become nigh impossible to cover. To the customer, rapid transits are as important as cost reductions, and Shafirkin (1978, pp. 125-6) shows just how significant this has become: his figures are reworked in Table 2.15 to show how much more quickly marchroute traffic moves over two distance bands.

Once again, the figures bring out clearly the high peak of performance in the period from 1965 to the early '70s. In no indicator is this revealed more clearly than in train weights and speeds, particularly the latter. Table 2.C shows the results in stark detail : after 1973, train speeds never returned to their peak, and in 1982 were 6.4% lower, although train weights continued their increase. The all-time peak of hourly freight train performance was 73,340 net tkm in 1973, the 1982 figure being 2% lower. Even more significantly, hours in movement of the average electric locomotive since 1970 have hovered between 9.6 and 9.7; while the same figure for diesels has declined from 10.2 to 9.4 in 1980.

Table 2.16 Freight cars - allocation of stationary time

	% time stationary	% of stationary time at			Yard time (hours) per 1000 car km	Absolute hours at terminals and way stations per car loaded
		terminals	way stations	yard		
1932	82.5	27.8	13.1	59.1	119.9	75.4
1940	82.4	30.5	13.5	56.0	79.1	64.1
1945	82.4	27.5	15.1	57.4	91.9	91.3
1950	82.0	35.0	15.0	50.0	67.2	73.7
1955	78.8	38.2	13.3	48.5	48.7	60.7
1960	76.7	41.6	12.9	45.5	36.9	56.1
1965	77.1	45.7	10.1	44.2	32.8	54.0
1970	77.1	43.4	11.2	45.5	32.9	56.3
1975	77.8	43.9	11.2	44.9	33.8	60.1
1980	78.3	(44.0)	11.7	(44.3)	(36.7)	(69.9)

Freight Car Stationary Time. No matter how efficient a railway may be, a freight car spends most of its time stationary. The SZD figures are much better than those of European and American railways, and are summarised in table 2.16.

Time at Yards. Railway statistics do not constitute an exact science, and Kozlov and Polikarpov (1981, p.224) show how approximate the collected data may be. Total freight car-hours in a period are calculated by multiplying the allocated cars by the period length; from this is subtracted hours in movement (freight car-km times train-hours divided by train-km). From that total, hours recorded at yards and terminals are deducted. The final result should, but never does, equal zero : Kozlov and Polikarpov comment 'as a rule, for a railway region the difference amounts to between ⅛% and 1⅛%', and it can of course be positive, or less probably, negative. The discrepancy is then pro-rated to recorded terminal hours and recorded yard hours, though only the marshalling element. In the example, and converting hours back to rolling stock, 36,429 cars were allocated, and the gap was 357, of which 237 were allocated to terminal and 120 to yards, on top of the 8903 reported there from the data from monthly form DO-24.

Soviet experience in yard operation is substantial, twelve times as much as in West Germany, for example. A most interesting survey appeared in ZhT (83/1/53-7), including a diagram relating detention time and throughput, from which Table 2.17 has been compiled.

Table 2.17 Statistics for main marshalling yards (1981)

Daily throughput	No. of yards	Total daily throughput (000)	Average hours detained	No. of yards with detention time more than 2 hours below average	above average
-2000	22	33.2	9.9	2	7
2000-4000	45	130.9	7.8	9	3
4000-6000	19	93.2	7.5	1	2
6000+	12	85.3	7.3	-	-
	98	342.6	7.8	12	12

The equation fitting the data for the last two columns is detention time = 28.05 times (daily throughput) $^{-0.158}$. The larger the negative exponent, the greater the economies of scale, and this operating statistic can be compared with the cost figures given by Sotnikov (1980, p.429) : operating cost per car per day at capacity is R 33.54 times (capacity/day) $^{-0.405}$, and capital cost R 557 times (capacity/day) $^{-0.471}$. (Larger yards have significantly lower costs, but are less able to reduce detention time.)

The article cross - analyses main yards by comparing 1981 and 1975 (see Table 2.18), the average detention time having increased from 8.8 to 9.9 hours :

Table 2.18   Changes in yard throughput and detention time (1975-81)

	Higher Throughput	Lower Throughput
Increased detention	45% (e.g. Kinel')	30% (e.g. Khabarovsk)
Decreased detention	13% (e.g. Minsk)	12% (e.g. L'vov)

In this table, Khabarovsk moved from 11.8 hours to 21.2, a figure so high that it could not be shown on the diagram. Ussuriysk is another named poor performer in the Far East, with 18.6 hours. 14 are shown with times under 6 hours, and 40 in the range 6-8 hours.

The top 100 yards carry out about 45% of all yard activity. For these yards, special returns show the distribution of hours under five headings, one of which covers transits without marshalling. Figures for 1965 and 1970 were published (ZhT 71/6/7), but the fact that these records are kept in such detail underlines the pressure for piecemeal planning made explicit by Sotnikov (1974, pp.359-68). There is a minimum detention for each yard at any one time, and a daily throughput minimising that detention time (in his example, 5.16 hours and 3750). Excess detention hours correspond roughly to [additional throughput/1250]^3; a yard cannot physically handle more than say 2000 above

the throughput at the minimum detention time.
Bearing in mind the need to keep the yard in
operation, planners have to avoid heavy investment
with long pay-off times on the one hand, and
frequent small investment on the other, since there
will be some disruption while the yard is upgraded.
As a consequence, Soviet practice has more and more
tended away from simple but largely implemented
projects such as more tracks, more switchers, more
humps, and more mechanisation, to parallel humping,
sub-yards and hump-avoiding tracks (some cars cannot
be humped e.g. those with explosives) and ways of
avoiding repeated humping (because cuts are
mis-shunted, couplers have failed to uncouple,
classification tracks are full, etc.).

Some yards must be close to their long-term
operating capacity, and every additional investment
becomes more expensive for an increasingly marginal
benefit. To take an obvious one, 64 classification
tracks are reckoned to be the upper limit, if only
because the outer tracks are more and more curved.
How does one explain to Gosplan that what is needed
is a brand new yard close by? Their answer would of
course be that more marchroutisation would postpone
a decision and therefore the expenditure. That
choice is in the hands of shippers, however.

Terminal Operations and Damage. Bureaucrats
frequently denounce delays at the 15,000 industrial
sidings, where about 90% of loading and 80% of
unloading takes place. In January 1982, Gudok
(82/1/31) reported a press conference on this topic
conducted by no less than the Deputy Minister.
Fines and even court cases are commonplace. An
example of the carrot rather than the stick is the
staging of 'socialist competitions' to break new
records in handling time. By Western standards, this
is not excessive - between 1955 and 1975 the lowest
figure was 42.2 hours (1959) per car turnround, and
the highest 47.9 (1975); there are about 2.15
handlings per turnround, reflecting trans-shipment
of LCL and other freight, and part discharges. While
it may seem that 8 hours is more than adequate to
load a car, that is not the end of the story; cars
have to wait for switchers to pick up and move them
on site, cars have to be cleaned, they may need to
be weighed after loading, and finally they probably
have to wait for a locomotive from the SZD to pick
them up again. Povorozhenko (1974, p.290) notes
that the first processes take 20-30% of terminal

44

time, and the last one 40-50%.

Sergeyev (1975, p.68) shows that terminal times are much lower for bulk goods : 21 hours for special cars with their unique facilities for freight handling, 29 for tank cars, 39 for open cars, but 74 hours for refrigerator and box cars. In Leningrad, where the port handled 12.2 million tons in 1980, it was reported that 70% more freight was handled in the first half of 1981 against the same period in 1980 by adopting 'integrated planning'; 6000 freight cars were released for other uses, idle time in the port reduced by 2.7 times, and the equivalent of two freighters of 10,000 DWT released. No doubt turnround time was still relatively high (in ports in the West it is two or three times higher than the average). How much of the 800 million tons handled by maritime and inland waterway shipping also involves rail is not known, but it must be at least half; at 45 tons per car, and excess demurrage of 2 days, some 50,000 cars would be tied up in ports.

ZhT (80/7/13-15) shows that although the number of car collisions on industrial sidings declined by a quarter between 1975 and 1979, the average cost per car repaired rose from R 47 to R 65. The article refers to 'tens of millions of expenditure', and it would seem that account 158 ('maintenance of freight cars in connection with loading/unloading') costs R 50-60 million a year. Total repairs are about R 570 million, or 8½% of the freight cost; the US figure, where the daily car movement is only a third as great, is about 7% of total cost. One reason why the Soviet repair cost is not higher is the encouragement given to shippers to make their own repairs with baling wire or anything else to hand. In Europe, about two thirds of the cars withdrawn from freight trains for treatment have body damage, while the rest require special treatment (brakes - 15%, wheels/axles/suspension - 10%, draw and buffing gear - 5%).

The problem, who repairs a freight car, has been one which European and North American railways have solved over a period of a hundred years. In Switzerland at any one time, 60% of rail-owned stock belongs to other countries. In the USSR in 1975, out of 3236 billion tkm, only 509 was internal to one of the 26 railways, and no less than 1270 billion was in transit across railways other than those loading or discharging. In the Soviet system the state owns everything and a railway does not possess its own car stock running under a common-user system such as EUROP in Western and OPW in Eastern Europe, to which the SZD in fact belongs : in 1981, Soviet rail exports were 2.1% and imports 0.5% of rail tonnage. Under such circumstances it has been well said (by the CPSU Central Committee) that the 'freight car has no owner'. Given a damaged car, a railway will foist it on to a hard-pressed shipper, or better still, transfer it to a neighbouring railway with a shortage of that type. Quite often it receives the same car later, still unrepaired. Consequently, regularising the system whereby shippers make running repairs has been favoured. A L'vov furniture factory (Gudok 82/12/14) repaired more than 1400 cars in 1982, and a paint factory near Yaroslavl' dealt with 1328. The Volgograd region of the Pri-Volga railway was even sending out repair specialists and inspectors and supplying spares, including couplings and brake parts. Since the average car is loaded and unloaded 120 times a year, and marshalled 240 times, it leads a dangerous life.

Line Capacity. Penny pinching occurs on the main lines as well as at the nodal points. In 1950, the average headway between electrically hauled freight trains was 45 km, and between steam/diesel 46 km. By 1965, this had reduced to 19 km and 42 km; in 1980 it was 14 km and 36 km respectively. These are averages and, of course, on many lines the figure is much less. Between 1960 and 1975, out of the R16,470 million spent on construction, 3170 million (19.3%) went on 12,500 km of new line, 2380 million (14.5%) on 9300 km of double tracking, 2110 million (12.8%) on 25,100 km of electrification, and 790 million (4.8%) for signalling and telecommunication : 16,400 km were converted to CTC and 37,700 km to automatic blocking, while 106 thousand switches were modified to electrical operation.

The money values may seen generous and the physical achievement colossal, but in reality the large annual traffic increases were only revealing new pressure points. Given the size of the country and its rail network, alternative routes rarely exist between any two points, as can be seen from some comparative density figures. Table 2.19 gives these for the Donetsk oblast, which is nearly as large as North Rhine Westfalia, and the Ruhr.

Kozlov and Polikarpov (1981) say that the most heavily used 20% of route carried 52.1% of traffic in 1940, 56.9% in 1960, and 57.3% in 1978. In that year, although a quarter of the route carried less than 4 million tkm/km, 9.2% carried more than 70

Table 2.19 Rail tonnage and route densities (Donetsk and Ruhr)

	Tons per day loaded per route km	Route km per 1000 km²
Donetsk (1975)	465	61
Ruhr (c. 1980)	550	350

million tkm/km, while the 1250 km at the top of the list carried an average of 141 million tkm/km. Multiplying these out, those intensively used sections of route carried 178 billion tkm, more than France, Germany, Italy, Switzerland and Austria combined.

In 1965, the density of traffic on double tracked routes was 4.57 times greater than that on single track, and trains moved 45% faster, partly because way stations on double track routes had been severely pruned. By 1974, the gap between the two had been reduced, as the figures from Mulyukin (1975) reveal (Table 2.20):

Table 2.20 Freight train speeds (1974); Time at way stations

	Freight train speeds km/hour		Time at way stations
	Technical	Yard to yard	
Single track	44.6	29.3	34.3%
Double track	49.5	39.7	19.8%

Between 1965 and 1975, the yard-to-yard speed varied from a low of 33.4 km/hour (1975) to a high of 33.9 in 1967, with 33.8 in each of 1968, 1971, and 1973. Sergeyev (1975, p.49) shows that over the 1966-70 period, the speed of trains on single track decreased by 0.2 km/hour and on double track by 1.5 km/hour. The reason why the overall average remained at about the same level is the growing proportion of traffic on double-track route (trains per day increased by 12.4% as against 2.6% on single track). By 1970, way stations were 33.8 km apart on double-track routes (15.3 km on single track).

The unit of measurement of line capacity can be gross or net tkm/pkm per km per year, freight and passenger cars (or axles) per day, or simply trains per day, and these are not interchangeable without a knowledge of train weights and consists. Passenger trains, including suburban, average 50 or more axles, and freight trains over 200, but Abramov makes it clear that consists on the most intensively used railways are nearly 20% greater than the system average. Making a train longer, without a corresponding increase in locomotive power, reduces its ability to brake and accelerate. To move over 150 trains a day in each direction on double-track route, as the 1250 route km mentioned above has to achieve, requires the most favoured conditions of closely spaced four-colour light signals, coupled with radio control, and absence of curves and gradients, and trains running at uniform speeds : most of all it needs to have a minimum of track failures, locomotive breakdowns, train derailments, and catenary disengagements. UIC leaflet 405R - Methods to be used for the Determination of the Capacity of Lines - shows in its 48 pages just how complex and important this subject is; readers are referred to the UN Study on East-West European Goods Traffic Flows (1979, pp.123-35, 152-7) for a comprehensive and readable analysis of the factors, and data on the thirty east-west through rail links.

Because of its critical importance, references to investment in increasing line capacity (propusknaya sposobnost') often appear in the literature; ZhT (77/2/74) states that double tracking permits another 32 million net tkm/km to pass, and CTC plus automatic blocking, 48 million on double track and 8 on single track. Since double tracking is about 12 times more costly than full CTC, the fact that CTC has not been applied to all lines carrying more than say 3 million net tkm/km must say something about its manufacturing availability and reliability. The cost, in terms of delays, of locomotive breakdowns must be substantial; there are perhaps 10 such each day. No doubt Gosplan's criticism on rail planning centres on the fact that 13% of double track electrified route carries less than 30 million tkm and pkm/km, and is therefore underutilised.

Economic Aspects of the Problem

The S Curve.   The S curve describes natural growth

48

phenomena where, because of competition or exhaustion, the growth rate gradually slows down and eventually flattens out. Does it apply to engineering artefacts such as a railway? Could it apply to the Soviet system? Some analysts have decribed developments such as the speed of passenger transport in terms of a series of S curves, giving the illusion of constant progress, although the technology moves in jumps.

The recent decline, both in train speeds and the hours of movement of locomotives as well as freight cars, indicates that this model does not now apply to the SZD. One can legitimately doubt whether the simple S curve concept can ever be applied to a railway per se; the development of 'progressive' traction might, however, be on one S curve, and general management efficiency on another. A better analytic model might indeed be traffic congestion, where for example peak flow is at half the free flow speed. The only indicator whose plot resembles an S curve is that of train weight (see table 2.C); the average annual increase from 1974 to 1982 was 16 gross tons compared with 50 from 1955 to 1974. At this rate, 1990 might see only 2920-2930 tons/train.

This prediction is flawed by evidence from the USA, where the gap between Class I railways and the SZD, which oscillated between 850-950 tonnes for a quarter of a century, has recently widened. In the mid '70s it was around the 1000 ton mark, and is now over 1100 tons. Heavily loaded lines in the US, like some of the Union Pacific main routes, must be exposed to similar pressures as the SZD; on average, the SZD carries five times as much traffic per route-km as the US. The pressures in the US for technological innovation are as great as in the USSR, but their introduction is far more successful. For all these reasons, the S curve does not seem to be a good descriptor or predictor of the benefits coming from railway technology in general and that in the USSR in particular.

Link between Industrial Production and Rail Traffic. Because the Soviet economy is the prime example of a smoke-stack economy, it is plausible to assume that rail traffic and industrial production are causally connected. Attempts to produce a rail-weighted production index in capitalist countries have failed, largely because of the increasing competition from other transport modes.

In the USSR, rail tonnages of iron and steel, and coal, are very similar to production figures, but output data to match most of the other rail commodity loadings are not available. Anyway, industrial diversification away from heavy inputs and outputs has taken place in the USSR too, so that linear equations (Y = A+BX) give poor correlations. Equations of the form Y = AX^E are better : only where E lies between 0.6 and 1 will these equations be acceptable to common sense, though they may still be usable; the black box represented by A and E is particularly opaque. Two sets of equations have been developed for our analyses, one with Y as tkm and X as industrial production in total, the other with Y as rail tons of individual commodities and X as production tonnages, where appropriate using an acceptable proxy.

Table 2.21 gives E for a variety of time spans, X being industrial production set at the base value of 1940 = 1000. The prediction errors shown against years after 1975, when the formula becomes more and more unreliable, come from the equation based on 1960-75), where Y = 25.704 x (industrial production)$^{0.6505}$. The inescapable conclusion is that the bad winter in 1979 (and probably 1982) had a lasting downward ratchet effect on rail traffic : the values of E for 1978, 1980, and 1981 (and perhaps 1983) support this hypothesis.

Table 2.21 Values of E to predict freight tkm (Y = AX^E)

1950-82	0.68	1975	0.61
1950-75	0.74	1976	0.37 (error-1.4%)
1950-60	0.83	1977	0.20 (error-3.8%)
1960-70	0.62	1978	0.65 (error-3.8%)
1970-80	0.56	1979	-0.66 (error-8.2%)
1965-70	0.60	1980	0.77 (error-7.8%)
1970-75	0.73	1981	0.56 (error-8.1%)
1975-80	0.28	1982	-0.35 (error-11.0%)

Table 2.D lists data by Economic Region over the period 1950 to 1975. Crude linear equations for annual changes in industrial production are available; information for industrial production by the 19 regions is not to hand for recent years. More fine-grain analysis has to depend on commodity data, given in Table 2.22; the equations are based on

outputs for 15 to 20 years in most cases.

Table 2.22 Values of E to predict rail tonnages by commodity from production ($Y = AX^E$)

Coal/coke	1.263	(coal)
Oil	0.395	(crude)
Ores	1.144	(iron/manganese)
Iron/steel	1.021	100% of finished steel plus 50% of crude steel
MBM	1.114	(cement)
Timber	2.103	(drevesina in cubic metres)
Grain	0.711	(procurement and imports)
Fertilizers	0.856	(mineral content)
Other	0.717	(sum of Group B and livestock products indexes : 1965 = 100)

This series overestimated total tonnage from 1970 to 1972, and undershot from 1973 to 1978 except for 1977; apart from grain, which depends on a 3 year moving average including imports, the overprediction was 28 million tons in 1979, 22 million (1980), 10 million (1981), and 44 million (1982). The series has the same downward ratchet effect as the previous one for 1979 and 1982, the effect being spread over most commodities. Both sets of equations have the same implication : the volume of rail traffic is not directly linked to industrial production, except in special commodities. Inputs to, and outputs from, the iron and steel industry account for 20% of total traffic, and do have such a link, for example. Significantly, in the '70s, the railway needed an annual traffic increase of 5% if its operating ratio was not to decline, and after 1975 only 1977 produced such an increase in industrial production.

The BAM.   Siberia, despite its inaccessibility, has long been considered by Gosplan as the El Dorado of the economy. The building of the Baykal-Amur Mainline is clearly one of the great speculative ventures of the century in transport construction; the mountainous terrain makes it the equivalent of about twenty Gotthard lines. It runs over 3000 km eastwards from Ust'-Kut, near Bratsk, the important site of hydroelectric power and an aluminium smelter. The line passes through the very

inhospitable country north of Lake Baykal via
Nizhneangarsk and the valleys of the upper Angara
and Chara to the Yakutsk ASSR border at Khani on the
Olekma. It then follows the Nyukzha valley to Larba,
and makes its way through Tynda (where it crosses
the line from Bam on the Trans-Siberian to Chul'man)
past the Zeya dam to Urgal. Its winding course then
follows the river Amgun to Komsomol'sk-na-Amure.

Its construction was partly to tap the region's
undoubted mineral deposits, partly to duplicate the
Trans-Siberian 300-400 km to its south and hence
well away from the Chinese frontier, and partly as
a prestige project long associated with Brezhnev.
Resources of national importance are found in most
of its nine sections : timber lies north west of
Lake Baykal, rare minerals in the Vitim valley,
non-ferrous metal ores in the Udokan mountains,
agrochemical minerals and coking coal in south
Yakutia, and coal has been produced for a decade in
Urgal. Timber is produced in the Amur valley; but
the ores most prominently mapped are copper, tin,
tungsten and gold, only the Bureya ridge seeming to
be without rare metallic or other mineral resources.

From an accounting point of view, the BAM seems
a marginal investment. The premium paid for railway
work in the area is at least 70%. The cost of double
tracking in the Urals and Siberia between 1976 and
1980 was R 680 thousand per kilometre, and in 1972
the construction of the line from Tyumen' to
Tobol'sk was R 699 thousand per kilometre. At a
modest R 2½ million/km to cope with preserving the
permafrost and providing earthquake protection, the
BAM investment would come to at least R 7½ billion,
plus rolling stock, at least 35% of the MPS
investment between 1976 and 1982. A generous
estimate of 30 million tkm/km, about three quarters
of the Trans-Baykal railway's current level, gives
annual revenue of about R 450 million; allowing 15
staff per km and a pay premium of 70% gives total
pay, at R 250 million, plus say 400 million for
amortisation and 150 million for other costs, making
a loss of about R 250 million. Kovrigin (1978,
p.116) gives data on railway variable costs to aid
our calculation of the BAM break even point.
Variable costs on the West Siberian were about R 690/
million tkm, the Northern 880, the East Siberian
800, and the Trans-Baykal 995; the figure of 1150
for the Far Eastern is distorted by the narrow-gauge
Sakhalin network. In current prices perhaps
R 500/million tkm is reasonable, given modern
technology, so traffic needs to be 50 million tkm/km

merely to break even : hence the concern over construction delays. All these figures rest on assumptions of optimistically low cost and high volumes.

Inefficiencies caused by the Number of Railways. In the early 1930s there were 22 railways, and after a post war increase to around 45, the number stabilised in the '60s at 25. In 1968, the Azerbaydzhan railway was split off from the Trans-Caucasus railway, exposing the latter's weaker position. More recently, the Kazakhstan railway, by far the longest, has been divided into three, and both the East and West Siberian railways have given birth, to the Krasnoyarsk and Kemerovo railways respectively. The Odessa-Kishinev railway has produced the Moldavian minnow, and the BAM has been set up as a railway in its own right.

Railway areas can never fully coincide with economic or political regions. At end 1974, 7.1% of Kazakhstan's route belonged to other railways, of which the main one was the South Urals. By contrast, 4.4% of the Kazakhstan railway's line lay outside the republic. Route length is only one of a number of factors determining the number of railways : in 1977, 13 were over 5000 km, 7 between 4000 and 5000, and 8 under 4000 km. Numerous reports concerning the 'hoarding' of empty freight cars and the 'theft' of locomotives from other railways show that self containment of traffic is one of the most important factors. Locomotives can go out to, and return back from exchange points, but freight cars cannot, unless they suffer trans-shipment, a very expensive activity. Together Shafirkin (1978, p.104) and Danilov (1977) give enough data to compile 1975 traffic exchanges for the 26 railways, shown in Table 2.23. Across the approximately 125 exchange points, there were about 2100 loaded cars per junction, all reported daily on form DO-1, specially designed for this purpose.

Tonnage handled (7384) can be divided by tonnage loaded (3621) to give a 'connected' ratio of 2.04, and this ratio used for international comparison. In 1950, the figure for US Class I railroads was 1.90, and by 1980 it had reduced to 1.63. In Britain in 1913, the ratio was 1.52 for coal/coke and 1.69 for general merchandise. These figures make two points clear. The first is that railways in a country which owe their existence to

Soviet Railways

Table 2.23 Freight movements across railways in 1975

	Tons (million)	Tkm (billion)	Daily loaded freight cars exchanged
Internal to one railway	1914	509	
Sent out ('export')	1704	800	110,000 out
Received ('import')	1663	657	110,000 in
Transit	2103	1270	155,000
	7384	3236	

economic reasons of organic growth have a lower connected ratio than if they are artificially created. The second is that smaller railways have higher ratios : for coal in Britain in 1913, the top four railways had 1.24, the next six 1.55, and the next ten 2.10, while in the USSR the Southern and South Eastern reached 3.9 in 1975.

The US and pre-war UK systems, as well as the RIV managed one in post-war Europe, worked on two main principles; each railway owned its vehicles, there being no common stock, and each railway receiving loaded cars paid a rent for them, usually on a per diem basis. Even then, there is the problem that net exporting railways are at a disadvantage vis-a-vis net importing railways, since the latter have the chance of reusing some of the cars received to load exports back to the owner. More complex arrangements have been developed in Europe to offset this bias, notably the EUROP and OPW agreements, to which nine West European and six (plus the USSR) East European countries are respectively party, which provides that the stock is virtually under common ownership; also, an importing railway which creates excessive empty movements may be required to pay for them.

The creation of more and more railways in the USSR, particularly where they are large net exporters (Donetsk, Kemerovo etc.) exacerbates the problem of controlling empty movements. In 1975, the Donetsk railway must have sent out each day 12 thousand loaded freight cars, and received 6000; another 6000 passed through in transit. In 1969, the average movement per loaded car was 1000 km, and excluding those reloaded on site, the average empty trip was 485 km. The Donetsk railway as a whole needed to import a new 6000 empties a day; Gundobin (1971, pp.520-4) demonstrates that most railways west of the Volga have to feed it with empties, at the rate of 6 for every 4 loaded open cars the Donetsk receives.

As an example of a railway with no economic viability, the Moldavian is hard to beat. Under pressure in 1940, Romania ceded Bessarabia, and in recent years the republic was one of Brezhnev's power centres. In the list of national groups by size in the USSR as a whole, Russians and Ukrainians come first and second by numbers, with Uzbeks third, Tatars sixth, and Moldavians tenth, with a 1979 population of 3 million. So, someone argued, they deserved their own railway. In 1975, the Moldavian (railway) region was the largest within the Odessa-Kishinev railway, itself ranked 17th out of the 26 in freight traffic and 20th in profit percentage on capital. Its transformation into a full railway means that the largest (October 10,040 km) is nine times the smallest (1160 km).

It is true that traffic growth since 1950 has been larger in percentage terms than in the USSR as a whole, but its volume is still small. The oblasts of either Kiev or Khar'kov, both with smaller areas, handle more tonnage. In terms of commodity movements, Moldavia is a net importer of everything except grain, fertilisers, and miscellaneous freight. Total tonnage handled in 1975 was 8.3 million tons internal, 7.5 million out, 20 million in, and an estimated 5 million in transit. A generous estimate of traffic would be 7.5 billion tkm, equal to 6.7 million tkm/km, the same as on the Pri-Baltic, the then lowest density railway. Freight costs must have been R 4250 to 4750/million tkm, in a year when average revenue was R 3971.

Together with its limited passenger revenue, the Moldavian railway must be one with a chronic deficit, which absorbs freight cars from other railways with little incentive to return them, and with full management resources for its diminutive length. Will the Central Asian or the Trans-Caucasus railways follow this example, spreading management thinner on the ground?

Soviet Railways

The Problem of Innovation

The Bureaucratic Process. Over most of the post war period, criticism of railway operation has concerned itself with questions such as 'how can costs be reduced further?' Occasionally, but much more frequently after 1975, the question has been broadened to 'are the railways fulfilling their allotted tasks?', and more and more since 1979 'have they been damaging the economy?'

Since there are no inner reserves left which can be exploited with ease, crucial decisions are required. First, can demands on the railway sector be reduced? To answer 'no' would be to enter the realm of the unwelcome and the unmentionable. However, if nothing is in fact done, that will be the result, unplanned and unwanted, and the railway will indeed be held to have damaged the economy. Second, there is the possibility of greater investment : more and bigger yards, more electrification, more bypasses, more double tracking and above all, more spares. Finally, there is 'new technology', to exploit the remaining inner reserves, particularly of line capacity.

On this battlefield Gosplan is usually for the new technology, the MPS usually in favour of more infrastructure. The latter's view is that what is needed is 'proportionality', by which it means a larger share of the investment cake. The crisis of 1979 provoked one Gosplanner to calculate that the railway's failures had choked off R 10½ billion of national demand, though perhaps this was only the same percentage of GDP as of traffic lost. Gosplan, particularly through its technical arm, the Institute of Complex Transport Problems, relies on induction: what worked in the past must work in the future. What is needed is a third revolution : it is easy to forget, because it is 50 years in the past, that the first could not have happened without more powerful locomotives together with such mundane things as improved couplings, brakes, and signalling.

ZhT (79/1/30-38) published a Gosplan inspired article, particularly interesting since it must have been written well in advance of the crisis, which said that as the key efficiency indicators were declining, what was needed was reconstruction rather than construction. Tables 2.24 and 2.25 continue the freight car and locomotive productivity data given for earlier periods in Tables 2.8 to 2.12. The results are less soundly based than the earlier ones, due to the decline in statistical data : in particular the indicators of static load and dynamic load seem to point in different directions.

Table 2.24   Freight car productivity factors (AARC) since 1965

	1965-75	1975-82
Traffic	+5.2%	+1.0%
Nos. of cars	+4.4%	+1.4%
Capacity/car	+0.8%	+0.6%
Utilisation	-0.3%	+0.4%
Loaded car-km %	+0.1%	N.A
Train speed	+0.3%	-0.9%
Hours in movement	-0.1%	-0.5%

Table 2.25   Locomotive productivity factors (AARC) since 1965

	1965-75	·1975-82
Numbers	+2.0%	+1.7%
Train weight	+1.6%	+0.7%
Hours in movement	+1.2%	-0.5%
Train speed	+0.3%	-0.9%

For Gosplan, reconstruction means a further intensification of traffic on existing lines. Since many of these are already carrying trains in excess of the 70-75% of capacity which engineers regard as the upper safe limit to be able to recover from breakdowns and bunching, this is simply to tighten the screws. Gosplan's dream solution required double-weight freight trains at 5-minute headways with a consist of 8-axle cars built to a new, larger, loading gauge. Such a solution, even if it were technically achievable, could not be brought in overnight, as no doubt the Ministry submitted. Someone sardonically pointed out that the wider freight cars would behead any worker standing between two passing trains.
    The real stumbling block in Gosplan's solution is that the new technology must be both available and capable of large scale application. The speed of the design process - conceptualisation, design,

test, acceptance, production, network wide diffusion, training - would have to be accelerated. Unfortunately, the railway is exposed to a Malthusian problem : capacity increases arithmetically while congestion grows geometrically; as congestion extends, so do signal checks, and the constant braking and acceleration increases locomotive stress; the resulting shortage of locomotives in good condition reduces train weights and speeds, and so the vicious circle is closed.

In theory, the railway is well provided with the technical institutes to produce the required innovations. At the bottom is the work coordinated by the All-Union Society of Inventors and Rationalisers, basically fed by suggestion schemes. Then there are the 15 institutes of railway engineers (VUZY) where 80% of those with doctorates in the railway service are employed : their main task, however, is teaching. It is the scientific research institutes (NII) who bear the main responsibility for developments, and the leader is the All-Union Scientific Research Institute of Railway Transport (VNIIZhT). This resulted from a wartime amalgamation, and is located at the advanced testing track 6 km long at Shcherbinka, 30 km south of Moscow. An affiliate in the Urals, which tackles problems jointly with the Sverdlovsk railway and one of the VUZY (the Urals Electromechanical Institute of Railway Transport Engineers), is also said to make important contributions.

Other ministries have research institutes, including those for car construction (VNIIV), and locomotives (VNIElI and VNITI). Since these lack testing facilities, there is ample scope for friction between supplier/industry institutes and VNIIZhT. Although fewer cases now occur of the latter sending drawings of unmakable or unfittable components and assemblies, no doubt the manufacturers still feel their hands are tied. The customer, of course, has the final say, after long or short tests. It is common experience in the West and East that urgently needed crucial components receive inadequate testing, while unwelcome 'improvements' are farmed out to the backwoods for extensive trials, in the hope that during the long delay a better alternative will turn up. ZhT (76/1/42-45) reported that the tests carried out by the Finns on the electric locomotives supplied by the USSR were the most rigorous that had been known, yet were of shorter duration than normal Soviet experience.

Before the war, the locomotive industry took over some of the railway ministry's research facilities, and today design work is firmly under the control of the supply industries. In any case, the reforms which brought in Brezhnev's production associations (1976) seem to have reorganised VNIElI and VNITI, with what effect is unclear.

On the occasions when criticism of VNIIZhT is published, it is very severe, and reminiscent of the thirties. Researchers are alleged to pursue their pet projects and neglect the important ones. Too many studies are of the 'a number of questions concerning the ...' type. In ZhT 80/6/2-4 in an unsigned article, which may unkindly have been published to mark the dismissal of the top echelon of VNIIZhT, the recent achievements of the Institute were listed. They were (1) the increase in permitted axle loads to 23 tons allowing an extra 80 million tons of traffic a year to be carried (2) the reduction of safety margins for loading timber (3) timetable preparation by computer. There then followed a long list of failures : backwardness in controlling the behaviour of hauled vehicles on track was singled out, presumably an oblique reference to the delayed introduction of the 8-axle car. After Andropov's remarks at the November 1982 Central Committee, ZhT (83/2/2-7) published an article bewailing the poor return on the billions of rubles spent. The slow application of computers was highlighted, particularly computerised timetabling! Other shortcomings included delays in developing new signalling systems as well as locomotive cab signalling, and failures to coordinate with industry, resulting in the delayed appearance of high-powered traction.

Outside observers usually blame slowness in innovation on the number of channels of communication, particularly horizontal ones. Where many organisations are involved, the failure of one component is lethal; usually the railway has to find a make-do and mend solution, as where the Sverdlovsk railway virtually rebuilds its diesel motive power (Gudok 83/4/8). Perhaps the real problem here is that the design process has no logical starting point.

Locomotive Development. Large-scale dieselisation got under way in the late fifties with the 4000 h.p. TE3, a somewhat unreliable performer but one which served its purpose. It was followed at the start of

59

the sixties by the 3000 h.p. TE10, and then by the twin section 2TE10L and 2TE10V, now the largest class in the USSR. It is odd that production of the TE3 continued until 1973, and that a new design (the 3000 h.p. M62) was also being built at Voroshilovgrad; perhaps the TE2's idiosyncrasies were at last being understood. Its failure rate in its early days was 24 per million km, but only half this in its last year of production; the 2TE10L had a failure rate of over 20, so that unreliability matched the increase in power. The M62 was designed to the European loading gauge and hence classified outside the TE system, and it borrowed many of the TE3's components. As an export model it may have benefited from not being designed by a committee. Its success probably increased Voroshilovgrad's self confidence.

Unfortunately, the 6000 h.p. twin section 2TE116 proved to be a failure. Following the M62, Voroshilovgrad had produced another 3000 h.p. export product, the TE109, with DC/AC transmission. Most railways have avoided or abandoned hydraulic transmission in favour of electric, but the DC system causes technical problems in matching power production and supply. The DC/AC system uses an alternator to generate electricity and a rectifier to convert it, and only solid-state technology can overcome the weight penalty. The TE114 for desert regions, and the 2TE116 were constructed as developments of the TE109. Three of the twin section 2TE116 locomotives were built in 1971, and 42 in 1972-3, after which annual output rose slowly to the hundreds. The locomotive had been controversial from the start, the Ministry opting for a new engine from Khar'kov, but Gosplan ruling in favour of a marine engine from Kolomna which the navy had rejected. The new locomotive went to Bukhara, Pechora, and Tyumen', and now seems to be concentrated on the Sverdlovsk railway.

It steadily became clear that the railway had to cope not merely with teething problems but with fundamentally unsound design. Acrimony moved up through the ministry hierarchy and many major components were redesigned. On the Sverdlovsk railway, according to Gudok (83/4/8), availability is between 48 and 60% and failures 80-90 per million km. In 1980, the locomotive was termed an 'unquestionable improvement', presumably after incorporation of the design changes. The industry castigated the MPS for its time consuming modifications devoid of improvement. VNIIZhT (ZhT

82/3/3-11) countered that it was an inadequate design on which too much time had already been wasted. Gosplan blamed both of them, while Voroshilovgrad continued to produce locomotives and spares to the original design. A real problem, intensified by the shortage of spares, is the overworking of locomotives. There are many lines where the loops are long enough to take heavier trains, but where trains have to be divided at the electric/diesel handover points.

Another possible solution was increasing axle loads up to 27 tons. Voroshilovgrad was to produce the 8000 h.p. 2TE121 and the 12,000 h.p. 2TE124 in bulk by the mid 1980s, as well as the 12,000 h.p. 4TE130 with 23-ton axle loads. The first of the latter went for trial on the BAM in August 1983, the same month as the 2TE121. Gosplan's rule of thumb, perhaps a spin-off from American experience, where only the Union Pacific went in for high powered diesel-electrics, seems to be 'more power equals more trouble'.

The shortage of high-powered electric locomotives is yet another common complaint. There is alleged to be 40,000 km of route where train weights could be increased by 2000 tons if only the power were available. Soviet locomotives, apart from the small classes of VL22m, VL11, and VL80c, are not easy to work in multiple. US railways have long averaged three locomotives per train, and South Africa has managed five at the head and six cut in the middle. (Chudov's textbook figures show 1.03 locomotives per freight train; of every ten additional locomotives assisting in a train, four were pushing, three double-heading, and three in multiple. Kozlov and Polikarpov do not make this detailed distinction five years later.) It is said that the VL22m pairs are so cumbersome to uncouple that it is easier to run them in pairs even when the power is not needed. The slowness of VNIIZhT in solving the technology of radio control of locomotives in the middle of the consist is to an outsider surprising. The Sverdlovsk railway found an ad hoc solution via synchronised brake applications, to reduce the chance of a train parting in the middle.

In 1960, only 10% of electrified route used AC, which lowers the copper requirement, and increases the distance between substations by $2\frac{1}{2}$ times; it also enables the railway to use electricity 'as transmitted', though it then needs rectification to DC on the locomotive. DC still accounts for over half the route length, although a kilometre of AC route carries, on average, 15% more traffic than DC: the top 20 AC substations each provide enough electricity annually for 16 billion gross tkm. Various European countries have developed specialisms such as multivoltage locomotives, multiple power units per axle, 50 KV and 3 phase supply, AC instead of DC motors, and the Swiss Re 6/6 has an incredible 1300 KW/axle : such leaps into the unknown are eschewed by the Soviets. It is planned to produce new, evolutionary, designs with 1000 KW/axle, and the BAM has been constructed with twin power lines (2 x 25 KV) simply to avoid the problems of either multiple voltage locomotives or the switch to a totally new voltage. One area where Soviet practice leads the world is catenary de-icing: large hydro-electric dams increase humidity in their neighbourhood, and since electrification takes place primarily to cope with heavy flows, the railway has special problems.

Freight and Passenger Car Development. In the 1970s axle loadings were increased first to 22 tons, then to 23-1/4, and 24 tons; recently 25 tons have been requested. This only results in the overloading of existing cars not designed for such loads. The fundamental problem is the new loading gauge which would require bridge strengthening, tunnel enlargement, and increased trackside clearances.

The saga of the 8-axle car is now a quarter of a century old. In 1982, ZhT reported that VNIIZhT was studying how much rebuilding of bridges with the standard 1884, 1896 and 1907 spans was needed. With these cars, net train weights could increase 1% annually, but there would be route availability restrictions. ZhT (80/6/70-71) revealed that some railways had made a start in locating such restrictive structures, while simultaneously other railways were constructing them. Gudok (80/6/13) implied that VNIIZhT's management was dismissed because it favoured the 6-axle rather than the 8-axle car. The plan to run exclusively 8-axle car trains on the BAM suggests that running stability where stock is mixed is a serious problem; similar difficulties emerged when British 4-axle container flat cars were run on the Continent.

The 8-axle car, should it ever appear in quantity, confers only one sizeable benefit - shorter trains for the same net tonnage. A 4000-ton (net) train would be about 530 metres long instead of 860. However, the cost per capacity ton is 10% greater, and Sergeyev (1975, pp.96-8) points to some practical operating consequences unpalatable to Gosplan. The reserve percentage and the empty-running percentage were greater, and dynamic tons per loaded axle were lower. Further indications of their possible use came from ZhT (79/5/56-58) : the planned flow in 'circular marchroute' was 250 million tons, which at the Sergeyev usage rate came to 50,000 cars; a third of them were on the East Siberian and the Tselinnaya, 14% on the West Siberian, and about 7% each on the Donetsk and Trans-Baykal; the remaining 9% was spread over four railways including the Northern and Far Eastern.

A much more important - and successful - innovation is the conversion to roller bearings. Soviet railways can afford far less than any other the delays caused by derailments and train fires brought on by molten axleboxes. Towards the end of the 1970s only 40% of the stock was so fitted; Dmitriyev (1980) notes that the investment pays for itself in 3 years. At the same time, hotbox detectors (PONAB) have been introduced, and about 1700 are now installed, stopping about 3000 trains a day at suitable points for checking; there is no information on the percentage of false alarms, the usual problem. Typically, their reports are not used to tighten up discipline at inspection points, so that some railways can report no benefits from PONAB. On the South Eastern, which occupies a key position between the Donetsk and Moscow railways as well as east-west, radio links between cab and detector have been installed.

Passenger traffic has benefited from the successful development of multiple units with 8, 10 and 14 cars on suburban traffic, which are large by Western standards. Diesel MUs are rarer and smaller, passenger costs being 2.4 times higher in 1975 than with EMUs. However, the most remarkable innovation, because it still has not emerged, is the couchette. Because of the distances travelled (over 30 million passengers a year travel on a journey of more than 2000 km), trains are composed of heavy sleepers with low passenger capacities : Soviet citizens are averse to tighter accommodation. ZhT (83/4/26-32) suggested that the raising of Soviet height restrictions would permit their introduction, an astonishing precondition for a European already accustomed to them. Perhaps here is one of those cases where the Soviet consumer has had his way. Shafirkin (1978, p.68) gives a number of examples

of long-distance traffic where rail increased its
share from 1970 to 1975 (Moscow to Sverdlovsk and
Baku, for example). Even to Irkutsk and Vladivostok
rail had 22% and 15% of the traffic. The customer
must feel he is getting value for money.

Future Developments

The current five-year plan.   1983 seems to have
recovered as well from the disaster of the previous
year as did 1980. Konarev, the new Minister, gave
his audience - which included a Politbureau member -
the usual line on the eve of the annual Railwaymen's
Day. The problems were due not to limited resources
but a lack of responsibility; shortfalls in the
first half of the eleventh five-year plan must be
made up in the second half by mobilising 'resources'
and 'scientific-technical progress' i.e. computers
etc.

It now appears that the 1985 target is 3880
billion   tkm   instead   of   the   original   3921,
according to the Deputy Head of Gosplan, Viktor
Biryukov. Taking the upper and lower plan figures
for commodities produces rail tonnages of 4252
million and 4359 million, using the formulae cited
earlier; Pavlovskiy's original projections were 4281
and 4387 million. All in all, it seems that the
railway must deliver 10% more than its maximum to
date.

Reducing the Demand for Rail Services.   The growing
economy causes an above average increase in the
movement of products with low loadability, which
fall into four groups. Between 1950 and 1970, the
first of these - LCL - was growing at the rate of
8.4% per annum, against 6.4% for all traffic. In
1970, it accounted for 0.86% of all traffic and 2½%
of loaded car-kilometres, and no doubt the growth
has continued. By 1975, a 1400 km haul took 270
hours, including terminal time, against 120 hours
for carload traffic. The second, container traffic,
increased at an even faster rate up to 1960, but
then slowed down; by 1975 it accounted for 2.05% of
freight   traffic   and   about   4%   of   loaded
car-kilometres. Its movement speed was little better
than LCL, 235 hours, mainly because of the absence
of large boxes (in 1980 80% of traffic was moved in
small containers averaging 3-ton capacity) and the
need for trans-shipment. The third category is

refrigerated traffic : on Sergeyev's figures this had 1.7% of traffic and 2.9% of loaded car-kilometres, but scattered indications are that its share of the total remains relatively static. The fourth category is that of bulky freight with low weight : the tariff has norms of 35 tons per 4-axle car for books, 33 for vodka, 18 for unprocessed rubbish, 13 for footwear, 11 for musical instruments, and 10 for haberdashery. This category accounts for about half of 'other' freight. All in all, these four groups require about 15% of loaded car-kilometres, and the proportion is growing. The opportunity cost of these services exceeds the accounting cost, and heavy freight such as coal has to show increased loadability merely to offset this effect. Arithmetically, this might be achieved if a small part of the production of new freight cars were of the 8-axle variety.

Very significant relief to the railway has come from the shift of crude oil, and to a much lesser extent oil products, to pipeline. It was not until 1974 that pipeline movement exceeded rail traffic in total oil movements; crude oil traffic by pipeline had overtaken rail traffic in 1961 and now accounts for 97% of tkm of pipeline movement. Since the revenue from moving crude oil by rail in 1975 was 4.09 times more expensive than by pipeline, the delay in laying such lines might be a fair complaint of the MPS; the density of crude oil movement by pipeline was 3.1 million tkm/km in 1960, 8.5 in 1970, and 20.1 in 1980, while in 1975 the average density of oil traffic on the Trans-Baykal railway was 10.8 million tkm/km.

Strategic Planning and its Benefits for the Railway. Three themes underlie the Soviet technical discussions on how the railway might provide a better service, which can be caricatured as the three Is - the mobilisation of Inner Reserves, the avoidance of Irrational Shipments, and the smoothing out of Irregular Flows. The first of these has been discussed in depth already; it is only necessary to add that there can also be inner reserves in line capacity. By 1979, for example, 87.8% of route had rail in excess of 50 kg/metre, average rail weight having increased since 1950 at about 1½% a year, the weight in 1937 being 35 kg/metre : 75 kg/metre rail is good for 600 million gross tons of traffic, 50 kg/metre for 350. In 1979, 54.9% of route had CTC and automatic

blocking, covering over 90% of movement.
    Productivity equations for the infrastructure
similar to those for rolling stock could be useful
indicators if only more detailed information were
available, for example subdividing double and single
track across electric and diesel haulage, and
perhaps CTC/automatic blocking from simpler
signalling. Such data would confirm or deny the
claims so often made for the capacity improvements
resulting from double tracking and modern
signalling. Away from the main lines, progress is
less remarkable i.e. inner reserves may be required
to make an even bigger relative contribution.
Overall, from 1960 to 1980, the number of switches
increased by 35%, the length of station tracks by
38%, and sorting tracks by 66% : freight traffic
increased by 129%. Many writers comment on the
shortage of siding tracks : ZhT (81/5/11-14) says
for example that the length should be 70-75% of
total route but is only 59.5%.
    The other two Is come from escaping from
customer imposed constraints, and are for that
reason preferred by the MPS. Irrational shipments
may only connote short hauls; in 1975, 10% of coal
tonnage moved less than 20 km (and 10% of ores, 2.8%
of cement and 2.3% of iron and steel). Less
frequently it means ultra long hauls : hauls over
3000 km for these four commodities were recorded for
4.4%, 3.9%, 0.9% and 10.9% of tonnage. Sometimes the
reference is to two-way flows of identical products.
It seems that 3.2% of iron and steel moves between
the Urals and West Siberia, 60% West-East and 40%
East-West, but how much of this is of truly
identical products is unknown.
    Between 1960 and 1979, when tonnage AARC was
+3.6%, hauls of under 50 km showed +3.44%; movements
of 3001-5000 km increased at 5.94% and 5001+ km at
5.83%. Since 1950, ALH has declined in only eight
years. Conventional wisdom is that this reflects a
highly desirable shift from rail to road, but in
1950 shipments moving less than 20 km amounted to
5.2%, and in 1979 4%, or 148 million tons. One such
study was reported by Tikhonchuk (1972) : in 1968
total rail tonnage was 2706 million, 76 million of
which was truck-rail, 368 million rail-truck, and 72
million truck-rail-truck. It was found that 71
million tons of bulk commodities, plus 19 million of
beet and peat and 25 million of other tonnage, were
transferable, accounting for 17% of freight hauled
less than 100 km, and 4% of freight in the range
100-200 km. The savings were equivalent to 30,000

freight cars out of 870,000, useful but not substantial, since speeding up trains by 2 km/hour would have produced the same effect. Loaded car-kilometres were reduced by ½%, and the analysis showed which railways would benefit. Unfortunately, the five most intensively used railways lost only 2.7% of their tonnage, while the five least intensively used lost 6.4%. A real problem for any customer was the fact that 70 million of the 115 million tons transferable to road was traffic moving from industrial siding to industrial siding, which would have required a large investment in trucks : a customer would be hard pressed to make such a transfer profitably.

Many short-hauls are effectively part of a production line process, as can be seen from the fact that they take several hours less than regression analysis predicts. Far more questionable are the ultra long-distance hauls, which are not dominated by military traffic, as is often thought: in 1975 225 million tons travelled 3000 km or more, and of that only 59 million was 'other freight'. The 30 million tons of coal travelling such a distance from the Kuzbass quadruples the pithead cost, but rough calculations show that an open car travelling 4000 km under load and 1500 km empty before its next load would clock up 18,000 tkm/day against the average of 10,000 tkm/day. More trips like that would do wonders for the operating indices, but with little value for the economy.

Irregular traffic flows have been ironed out, at least as far as freight is concerned. In 1967, the two morning peak hours accounted for 22.6% of suburban passengers leaving stations, and 39% of inter-city passengers travel between 1 June and 31 August. In the West, railways suffer from intensive peak days, Germany for example loading 22% of freight cars on a Friday; only half of German tonnage comes from points open 'round the clock'. Shafirkin (1978, p.117) enables us to calculate the coefficient of variation for monthly loadings. In 1950 it was 12%, but only 4½% in 1960, and 3½% by 1975. A similar reduction applies across all commodities apart from timber. Shafirkin also gives data on relative loadings on Sundays, Mondays, holidays and return-to-work days : loading on these days is as high as 96.2% of the year's average day. (From 1963 to 1982, in Czechoslovakia, Sunday loadings were 81.5% of the average day, and unloadings 94.0%.)

Railways usually have the problem of the

geographical unevenness of commodities. In Poland in 1980, 49.7% of tonnage was loaded in Katowice, one out of the 49 voivodships. The USSR has a similar unevenness of coal, ores, timber, and oil, but rarely do these coincide, and the dispersion of building materials over the whole country provides a welcome counter weight. Perevozki gruzov (1972, p.169) shows that in 1965, 4760 stations loaded MBM, 1028 with over 100,000 tons each. In 1975, 6900 tons were loaded per average route-kilometre but only 5 railways had either less than 5000, or more than 10,000 tons/km. Another benefit clearly perceived by management is the greater probability of keeping an MBM-loaded car within their territory; bricks have an ALH of only 350 km, for example.

Overall, the major towns are net receivers of freight, Moscow oblast sending only half as much as it takes, and coal areas are net despatchers, Kemerovo oblast receiving half as much as it loads. The tonnage data for 126 oblasts and republics, generally available up to 1972, enable the changing balance to be measured. The hypothesis is that net receivers show higher growth rates in loadings, and net senders in unloadings; Table 2.26 demonstrates that this hypothesis is supported by the data between 1960 and 1970, and other more limited analyses confirm it over other periods.

Table 2.26 AARC in area tonnages loaded and unloaded (1960-70)

1960 - loaded tonnage as % of tonnage loaded + unloaded	AARC	
	loaded	unloaded
-34.9%	+5.65%	+4.11%
35-44.9%	+5.22%	+4.88%
45-54.9%	+4.55%	+3.72%
55-64.9%	+3.51%	+4.11%
65%+	+4.09%	+5.76%
All	+4.39%	+4.39%

Computer Management and the New Technology. The new technology has long been heralded as spearheading the third revolution. Signalling and telecommunications is one area where this should have been visible to the outsider, especially since so much of the route is now covered by CTC. In 1940, per 1000 route km there were 535 S & T staff, in 1950 619, in 1970 804 and in 1975 928; only in 1980 did this drop back to 850. Abramov (1974, p.89) gives a diagram relating S & T expense in 1972 to annual trains including light-running locomotives, both expressed per route km. (The busiest railway had about 33,300 trains a year, 91 a day.) The cost equation was annual S & T rubles = 630 + 23 times daily movements, with a correlation coefficient of 0.876; of course, the busiest railways have some low density route, and vice versa. New technology seems to have had little effect on costs and staffing in the S & T division.

The volume of railway work which can be computerised is vast. The West Germans spend ½% of total costs on such work, with 300 application areas, 32 thousand programs, and 21 million instructions. The most exciting projects are those which control freight car placing and minimise empty running, but these require an enormous input volume, much of which is uncheckable. It also requires all the links to be computerised, particularly train consists. According to ZhT (84/2/18-21) only 20% of these are currently computer processed, and the need for accurate input is underlined by the article, which has the title 'New Freight Car Numbering System'. The existing system was introduced in 1963 at the same time as a pan-European system was adopted under UIC auspices : the Soviet system used the same number of digits (7) and the same meaning for the first digit i.e. type of freight car. The new SZD system adopts the same check-digit system as the UIC adopted over 20 years ago for the eighth and last number. The logistics of managing the changeover are frightening, since every depot will have to have a one-ton-one cross reference table for the 1.9 million cars in the total park, and car reports must allow both systems.

Such criticism as has been made has been that such systems require accurate input, not always forthcoming from brawny railwaymen. The West Germans introduced a unified numbering system (complete with check digit) for all rolling stock including locomotives as far back as 1 January 1968. Is the delay in doing so in the USSR another example of the difficulty in making innovations?

Both Khandkarov (1977) and Tsarev (1981) give some idea of the work being entrusted to computers, though it is impossible to tell how extensive the applications are. By input point, the analysis is

30% for freight accounts, 20% depots, 15% main
sorting yards, freight stations and other
enterprises 10% each, and 5% for each of passenger
stations and minor yards. A table of applications in
Khandkarov gave the October and Moscow railways the
largest number and the Azerbaydzhan and Far Eastern
the smallest. 15% of work is on a 'real-time scale'
(4-6 hours), and 20% monthly or less frequently.
Table 2.27 reproduces one in Tsarev (1981, p.252).

Table 2.27   Computer Systems in late 1970s

ASUD (rolling stock) 30%	ASZhS (statistics)	12%
ASPRZhT (planning/	ASUMTO (materials)	3%
control) 10%	ASU Fin	1%
ASUM (freight) 16%	ASBU (accounting)	2%
ASUT (locomotives) 6%	ASUL (passenger)	6%
ASUV (passenger/	ASU Kadry (personnel)	2%
freight cars 5%	ASNTI (OR)	1%
ASUE (electrification) 3%		
ASUP (track) 2%		
ASUKS (construction) 12%		

Perhaps, after all, the Belorussian railway pilot
computer project reported in ZhT (82/8/5-11) has
been a resounding success, and points the way to
future successful developments.

Railway Management.   The quasi-military structure
of railways - on British Rail senior staff are
called 'officers' - is a precondition of success.
There is much speculation as to whether the SZD is
over- or under-managed. A major problem is that
railways have to be organised in a three dimensional
cut, by function, business sector, and geography. In
any one area, freight management has a claim on
shared functional resources such as track and
rolling stock, much of which moves between areas.
British Rail, with one eightieth of the SZD's
traffic, has 5 business sectors, 5 regions, 18
signals and telecomms divisions, 20 traffic
divisions, 27 civil engineering divisions, and 88
rolling stock depots.
    European railways have between 10% and 15% of
staff on administration, a proportion which has
grown over time. There are diseconomies of scale;

the larger countries have about 70 such staff per million train-km, the smaller ones 40. Kovrigin (1978, p.118) gives an interesting breakdown of management by the ten or so functions, in which it appears that 13.4% of total cost is for divisional management, the largest blocks being locomotives, track, and freight and passenger rolling stock; 1.6% is for regional management and 1.8% for the 32 railway managements. As in Italy, an unknown number of staff are tucked away in government departments; Kozlòv and Polikarpov (1981, p.258) make it clear that another 220,000 are employed in rail medical institutions, 200,000 in education (from kindergartens to colleges for cadres), and 260,000 in provisioning organisations.

No doubt the construction of the BAM and the fission of railways such as the Kazakhstan has absorbed some railway management talent, but it seems unlikely that this drain will have proved excessive. It is impossible to record a clear verdict on the quantity and quality of rail management.

Labour Morale. Improvements in labour attitudes may well be the factor to give most benefit, but they are not easy to measure. Under the Brezhnev regime, cadres became more secure and labour sanctions less effective, if only because pay rises in the absence of goods to spend them on become increasingly valueless. In these circumstances the sacking of top officials must have been seen as undeserved bad luck by those who kept their jobs. The minister, Pavlovskiy, was dismissed on 22 November 1982, followed later by the heads of the Alma Ata and Trans-Baykal railways, ranked fifth and tenth in terms of labour productivity in 1978.

Indiscipline and irresponsibility are the charges raised incessantly against railway workers. The combination of traditional Russian inertia and a couldn't-care-less attitude seems to cause incredible mishaps. Gudok (83/1/20) reported a derailed postal car bouncing along 9 kilometres of sleepers on the main line near Kuybyshev; two railwaymen were inside it, but neither took any action because one was too busy completing his official forms and the other was 'off duty'. Another issue referred to a locomotive driver who kept his job after being found guilty of exceeding speed limits, passing stop signals, and tampering with his speed recorder. In September 1982, Gudok singled out 38 locomotive depots, perhaps 8% of the total, where it was commonplace for crews to be drunk or asleep, not to mention the case of the citizen who housed her piglets in a railway container.

Managerial indifference is probably even more deep-seated and damaging. The range of sticks and carrots which are permissible is limited and already used to excess. From time to time investigations turn up results to surprise even the cadres. The Gor'kiy railway, according to ZhT (81/9/59), set up a study on the causes of high labour turnover, which blamed poor housing. In ZhT (81/5/27), the question was, why is train despatching considered a low grade job? The answer was low pay. Gudok from time to time refers to similar surveys, some of which have been abruptly terminated.

It must be noted that railway pay has been increasing relative to the national average ever since 1971; in 1982 the differential was 11.7%. Even more importantly, the premium for work in environmentally difficult areas has increased. In 1974, the Trans-Baykal paid 25% premium to the average worker, the Far Eastern 24%, the Northern 18%, the East Siberian 16%; railways paying premiums between 8% and 4% were the Sverdlovsk, the West Siberian, the Kazakhstan, the South Urals and the October. These railways employed 820,000 people out of the 2.06 million total, and 620,000 were in or east of the Urals. These differentials were of course on top of the extra then paid to railway workers as a whole, R 1873 a year against the national average of R 1693.

Locomotive crews were the best paid, with 30% more than the rail average. Track staff got 6% less and signals and telecommunications staff 8% less, while station staff got 24% less. The employees suffering low relative pay were those whose limited skills were most transferable, and vice versa. Some attempt has been made to upgrade staff : Tsarev (1981, p.101) shows that between the years 1966 and 1979 the top two grades accounted for an additional 167,000 staff, 403,000 in all. The proportion of freight station staff in the bottom two grades fell from 74% to 60%. The average grade went from 2.9 to 3.4 in the 6-point scale, and the resulting wage drift was worth about +½% per year.

Railwaymen are, in principle, stayers : Shafirkin (1971, p.15) shows that in 1967 30% of railway workers had less than 10 years' service, as against 40.8% in the overall economy, and 30.5% had more than 20 years, as against 20.7%. The average age was 38.3 years, more than two years older than the economy-wide average. 24.5% of rail staff, versus 19.3% in the national economy, were aged between 45 and 60, whereas below 25 the figures were 9.9% and 15.8%. Perhaps the rail system is caught between the conservatism of middle age and irresponsibility of youth, who have little to spend their money on and even less in the way of sanctions from their elders.

New Control Indicators. In December 1982, the Central Committee switched the main measure of output from traffic to tonnage shipped. Ton-kilometres, as an index, flatters long hauls and railways with transit traffic, whereas tonnage loaded is thought to be a better measurement of management performance. However, if railways use this change of emphasis to load empty freight cars, railways like the Donetsk, Pri-Dnepr, Kemerovo, Krasnoyarsk, and Kuybyshev will be in real trouble : these five load 35% of outward tonnage and unload 20% of inward tonnage excluding transit. The major transit railways such as the Kuybyshev, Gor'kiy, South Urals, Sverdlovsk and South Eastern, which take three of every eight transit tkm, will downgrade such movements. All in all, one must ask if the Soviets are returning to measuring chandelier production in tons.

One of the important input indicators, freight car turnround time, is enjoying a new lease of life. The index for an individual railway cannot, unfortunately, be calculated in the same way as it is for the SZD as an entity. The calculation is turn round time (days) = allocated cars x days in the period divided by cars loaded plus those received loaded for local unloading or in transit : applied over all 32 railways this index is roughly half the real figure. Gosplan has taken over from the MPS the responsibility for target setting, and will set them with greater stringency, hoping thereby to get quicker turnrounds at terminals and less delay in yards, as well as faster trains. Railways in their turn want some credit for handling empties, which, generally speaking, have longer stays in yards and, when in train loads, have lower priorities.

All the signs are that Gosplan is in search of the holy grail of one general purpose indicator, and will be baffled when its planners do not find it. The importance of synergy in railway operations seems not to be appreciated, and the fact that all the indicators are loosely interconnected will bedevil their use of a limited number for control purposes. So too will the absence of quality indicators such as the coefficient of variation round transit times for a given distance, which is about 40%, to judge from Kozlov and Polikarpov.

The Urals - Point of Maximum Vulnerability

The Urals can be defined as either the economic region, into which the Bashkir ASSR intrudes, or the area covered by the Sverdlovsk and the South Urals railways, or some larger grouping such as Dewdney's. An area of about 1 million square km covers the economic region plus the Bashkir and Tatar ASSRs and Kuybyshev oblast : in 1972, the 14,000 rail km of this area loaded just under 14% of USSR tonnage. In 1975, the two railways accounted for 12.3% of tkm and 6.9% of pkm, average freight loading being 38.2 million tkm/km; the overall load on electrified line on the South Urals was as high as 81 million tkm/km.

The South Urals railway has long been at the head of the league table. In 1974, its profit was R 90,300 per route-km, against a national average of 44,700. Its freight costs were 76.7% of the average, only the West Siberian having a lower figure. In terms of labour productivity it is also at the top, as Table 2.28 demonstrates.

Table 2.28 Labour productivity (thousand tkm/pkm per employee) on selected railways in 1975

		Freight			Passenger	Combined
	All traction	Electric	Diesel	Steam	All traction	
South Urals	3343	3831	2282	875	685	2870
West Siberian	2990	3198	2458	633	650	2506
Gor'kiy	2559	3451	2169	985	748	2145
Kuybyshev	2553	2921	1825	1090	933	2265
Sverdlovsk	2362	2414	2311	559	597	1975
SZD overall	2032	2500	1737	454	657	1715

Apart from its position as a leading producer of iron and steel, two other factors make the

74

railways in the Urals of central importance. The
first is the hilly terrain, which impeded other
forms of transport. The second is the limited number
of through east-west routes. To the west there are
five over an 800 km north-south extent, through
Perm', Sarapul, Ufa, Orenburg and Ural'sk. To the
east, even now there are only three, over a 400 km
extent, via Omsk, Irtyshsk and Pavlodar.

The area has been a centre of heavy industry
since before the revolution, but it was the
investment during the 1930s which gave it its most
substantial development. Table 2.29 gives the
shares of total rail tonnage in 1937 and 1975; in
1937 the area was defined as that covered by the
Sverdlovsk, Chelyabinsk, Bashkir, and Orenburg
railways, and the total tonnage in the USSR 517.3
million, while in 1975 it covered the economic
region only, and the total tonnage was 3621 million.

Table 2.29   Urals share of total rail tonnage

	1937	1975
Internal	5.7%	5.5%
Urals to west	1.9%	3.0%
Urals to east	1.0%	1.4%
West to Urals	1.1%	1.8%
East to Urals	1.8%	2.8%
Transit west-east	1.3%	1.7%
Transit east-west	1.3%	2.8%
Total	14.1%	19.0%

The flows on the western side in 1975 totalled 334
million tons, and on the eastern, 311 million.

While weather conditions in the Urals are not
the worst in the USSR, the January conditions of
-15° Celsius and 15 mm of precipitation are fairly
bleak. In the bad January of 1982, Perm' had 46,
Sverdlovsk 37, Tobol'sk 69, and Orenburg 58 mm of
precipitation. ZhT was quoted above on the problems
posed by the 1979 winter on yards; Problemy
razvitiya SSSR (1981, p.61) gives useful data on
tonnages handled by nodes, summarised in Table 2.30.
The major nodes in the Urals area probably include
Perm', Sverdlovsk, Orsk, Kartaly, Chelyabinsk,
Kustanay, Tyumen', Serov, Kurgan, Chusovskaya,
Yegorshino, and Kamensk Ural'skiy : excluded are the
exchange yards at Kinel', Sol'-Iletsk, Nikel'tau,
Tobol, Troitsk, Novo Uritskoye, Volodarskoye,

Soviet Railways

Table 2.30 Rail tonnage in traffic nodes (number of nodes)

	1975		1985 forecast	
	USSR	Urals/ W.Siberia	USSR	Urals/ W.Siberia
- 50 million tons	35	4	16	4
51-100 " "	46	5	36	3
101-200 " "	42	12	41	2
201-300 " "	1	-	32	5
301-500 " "	-	-	28	8
501+ " "	-	-	4	3
Not specified	-	-	9	1
Total	125	21	166	26

Petropavlovsk, Isil'-kul', Kolchedan, Polevskoye, Mikhailovskiye and Kropachevo on the South Urals, and Druzhinino, Piban'shur and Nazivayevskaya on the Sverdlovsk. The yards at Perm', Sverdlovsk and Chelyabinsk are often referred to, and must be amongst the largest and most developed in the USSR, that at Chelyabinsk showing a 'profit' of 19% in 1969.

ZhT (79/7/56) gives plenty of data to expand our knowledge of major routes in 1977. Four lines were examined in detail and the results are listed in Table 2.31. Plausibly, each of these sections had loads of the order of 100 million tkm/km, and trains at least as frequently as one every 10 minutes. One can conclude from the above data that the heaviest load is along the Kartaly-Chelyabinsk-Kropachevo line, demonstrating the need for the Magnitogorsk-Ufa extension and the Ufa by-pass.

Over an 11-year period, the South Urals increase in traffic exceeded the SZD average in six; in 1972 it was 9.2% more than 1971, against 4.7% for the network as a whole. Can the region take further substantial increases? No doubt the MPS and Gosplan wish they could give unequivocal answers.

Postscript
The problems of 1982 seem, in retrospect, to have been more than successfully overcome: the first three months of 1983 produced freight traffic of 886

Table 2.31   Trains on major sections in the Urals

	km	Length below norm	Length at norm	Weight at norm, length below	Weight against norm under 70%	over 90%
Petropavlovsk-Chelyabinsk	525	7.4%	64.3%	28.3%	25.4%	37.1%
reverse direction		17.0%	67.1%	15.9%	46.2%	39.7%
Chelyabinsk-Kartaly	261	7.4%	67.2%	25.4%	17.5%	41.0%
reverse direction		13.8%	33.6%	52.6%	15.9%	72.2%
Chelyabinsk-Kropachevo	264	9.9%	39.5%	47.6%	13.1%	69.3%
reverse direction		3.4%	66.9%	29.7%	34.2%	45.9%
Sverdlovsk-Balezino	557	4.8%	65.0%	30.2%	9.7%	57.6%
reverse direction		0.4%	76.5%	23.1%	20.8%	60.6%

billion tkm, 5.4% more than in the same period in 1982, but more newsworthy still was the fact that the average daily rate exceeded 10 billion tkm for the first time ever in the second quarter of 1983. 1984 looks as if it will exceed 1983's performance by some 2%, to judge from the figures for the first half of that year.

This must have been achieved in association with rolling stock utilisation rates returning to but not exceeding previous levels. The league tables published in Gudok fom time to time for the 32 railways seem to have reached a frenetic pitch. Thus in August 1984 average freight train weight on the Baykal-Amur railway exceeded plan by 299 tons, followed by the Kemerovo railway with 156 tons. The 1983 average freight train weight increase of 30 tons was much better than in recent years, and comparable to the increases of the early 70s, but the increase of 85 tons in the first half of 1984 was much more impressive. However, it implies that the much vaunted operation of super-giant trains is still limited. The Tselinnaya railway, often mentioned in this connection, achieved +118 tons against plan in August 1984, and +126 in September;

it was beaten by six railways in August, and four in
September. Presumably the publicity is intended to
impress, in particular, Gosplan.

The stresses under which the railways between
the Urals and Lake Baykal operate still remain.
Table 2.32 shows a selection of statistics for these
railways in league table form (1 being the best
railway and 32 the worst); the tables are in Gudok
(84/9/7, 84/9/5, 84/7/3). In freight train operation
the Moldavian railway ran closest to schedule,
followed by the Baykal-Amur, a reflection (no doubt)
on their lack of intensity of use.

Table 2.32   Railway league table positions against
plan (stated months in 1984)

	Running to Schedule (August)		Train weight (August)	Track renewal rate (June)
	Freight	Passenger		
Sverdlovsk	30	13	30	3
South Urals	24	10	24	11
West Siberian	21	19	16	1
Kemorovo	32	26	2	22
Krasnoyarsk	13	5	12	9
East Siberian	18	17	18	7

As a proxy for overall performance, the league
table positions for a variety of indicators can be
added together, and the railways then ranked on
this composite total. This operation on the three
factors in Table 2.32, excluding passenger train
performance, shows that the Belorussian was by far
the best railway: the next four best were the South
Western, Far Eastern, Krasnoyarsk, and Northern
railways. The Sverdlovsk railway came in at 27 and
the South Urals at 22: the four Siberian railways,
in the same order as in Table 2.32, were at
positions 8, 20, 4 and 11. Lower than the Sverdlovsk
were the Gor'kiy, Trans-Caucasus, North Caucasus,
and Azerbaydzhan; the Baykal-Amur is excluded
because of course it had no track renewal work.

It does seem that significant improvement in
one indicator is often associated with marginal
improvement in others (Sverdlovsk, Kemerovo). The
appreciative overall gains in productivity needed
from locomotive departments, and from track layout

and maintenance, and signalling, still seem to be a long way from achievement. Every little counts, however, and it seems that this will have to suffice until a major technological breakthrough.

Soviet Railways

Table 2.A RAILWAY FINANCES (1975 and other years)

	Share	Gross profit Relative per km	Gross profit % on capital 1975	Gross profit % on capital 1969	1968 % of revenue	Amortisation R/km	Amortisation R/mn tkm	Local revenue Share (1973)
October	4.1%	563	7.1	10.6	38.2	18040	999	38.2%
Pri-Baltic	0.9%	202	3.4	7.1	29.9	14320	1512	36.9%
Belorussian	2.1%	527	7.1	8.7	30.1	16610	1116	32.0%
Moscow	8.1%	1212	10.4	12.2	36.7	27210	907	27.2%
Gor'kiy	6.4%	1584	16.3	20.5	52.9	23980	601	16.0%
Northern	3.0%	734	8.0	13.1	37.1	20700	721	28.5%
South Western	3.3%	989	11.0	13.9	42.6	21110	795	17.2%
L'vov	1.2%	359	4.4	7.3	27.0	17920	1238	27.3%
Odessa-Kishinev	2.3%	620	6.7	12.1	42.5	21640	1033	21.9%
Southern	2.7%	1052	9.6	13.8	41.1	26470	870	12.9%
Donetsk	3.4%	1627	11.2	15.9	44.3	34770	930	32.6%
Pri-Dnepr	3.0%	1271	11.3	14.3	44.8	25490	795	25.2%
North Caucasus	5.6%	1409	13.5	18.1	49.1	24550	883	28.8%
Azerbaydzhan	0.9%	633	6.6	14.0	48.8	22050	1058	28.3%
Trans-Caucasus	0.3%	234	1.9	5.0	21.5	23010	1943	56.8%
South Eastern	3.7%	1368	13.0	14.8	40.0	26090	682	14.0%
Kuybyshev	6.2%	1884	15.5	20.3	51.2	30550	667	15.5%
Pri-Volga	3.8%	1182	12.8	18.1	49.3	21750	749	18.1%
Kazakhstan	6.9%	694	10.3	16.7	49.5	16680	654	25.0%
Central Asia	1.5%	348	5.3	11.7	42.2	16550	1069	59.6%
Sverdlovsk	5.7%	1382	13.2	19.8	49.0	24380	708	25.0%
South Urals	7.2%	2070	17.4	19.8	48.9	27420	481	13.7%

West Siberia	8.4%	2002	17.0	20.7	50.5	27300	532	22.2%
East Siberia	5.4%	1334	9.4	10.8	40.5	29460	729	19.1%
Trans-Baykal	2.8%	1139	9.0	13.8	43.7	27060	676	11.4%
Far Eastern	1.1%	331	2.6	5.2	23.8	24330	1145	41.2%
Total	100.0%	1000	10.3	14.2	43.6	22580	781	24.4%

Gross profit = Local revenue + shared revenue - costs (excluding interest)
Local revenue = Local freight + supplementary revenue + local inter city + suburban
        (Silayev 1975, pp.37-8)

The amortisation cost for freight movement is estimated from an arbitrary allocation;
the passenger share of amortisation is higher in absolute terms for the traffic carried,
but lower as a percentage of total cost.

Table 2.B KEY PRODUCTIVITY INDICATORS BY YEARS

| | Per freight car (4 axle equivalent) | | | | Per freight loco | | Freight trains in movement |
| | Daily net tkm | per capacity tonne | Daily hours in movement | Daily km | Daily net tkm (000) | tkm | Net tkm/hour |
					Electric	Diesel	All traction types
1950	3750	91	4.32	146.4	220	147	27550
1955	5040	106	5.09	188.2	370	296	37170
1960	6310	121	5.60	227.0	621	594	48520
1965	7250	129	5.49	249.7	727	654	60340
1966	7310	129	5.43	247.9	734	643	61470
1967	7470	130	5.43	250.9	763	638	63760
1968	7530	129	5.40	250.3	766	625	65280
1969	7580	128	5.37	249.3	778	615	66670
1970	7780	129	5.49	255.5	762	650	67840
1971	7930	132	5.54	258.2	784	614	68970
1972	7890	132	5.47	254.8	785	609	69900
1973	7980	133	5.46	255.2	778	630	71230
1974	8080	134	5.48	257.1	777	649	71960
1975	7890	130	5.32	248.5	780	666	72840
1976	7780	127	5.28	244.5	753	657	73340
1977	7530	122	5.12	234.5	745	650	72820
1978	7580	122	5.19	233.9	745	658	72390
1979	7310	117	5.09	223.9	729	634	71170
1980	7520	119	5.21	227.0	729	632	71160
1981	7740	123	5.28	232.2			(72130)
1982	7700	120	5.12	223.7			(71890)

Table 2.C FREIGHT TRAIN STATISTICS BY YEARS

	Freight Train Weight(T)		Freight Train Speed (km/hour)			Freight Train Locomotives	
	Gross	Net	Technical	of which electric	Yard-yard	Daily km	of which electric
1950	1430	815	33.8	39.6	20.1		
1955	1758	1002	37.1	41.8	24.7		
1960	2099	1201	40.4	46.5	28.3		
1965	2368	1332	45.3	50.1	33.6	390	500
1966	2406	1348	45.6	50.1	33.7		
1967	2460	1383	46.1	50.3	33.9		
1968	2497	1410	46.3	50.2	33.8		
1969	2537	1440	46.3	49.8	33.5		
1970	2574	1462	46.4	49.6	33.5	457.8	502.0
1971	2597		46.6		33.8		
1972	2631		46.6		33.7		
1973	2675		46.7		33.8		
1974	2708		46.6		33.5		
1975	2732	1563	46.6	48.4	33.4	465.3	490.5
1976	2741		46.3		32.9		
1977			45.8		32.3		
1978	2777	1605	45.1	45.6	32.1		
1979			44.0		31.0	428.5	438.7
1980	2819	1632	43.6	43.1	30.6	425.1	434.3
1981			44.0		30.9	426.1	441.0
1982	2839		43.7		30.6	420.8	441.2
1983	2870						

Note : Tables 2.B and 2.C   Sources are various, mainly Transport i svyaz' (1972), Kozlov and Polikarpov (1981), Silayev (1975), ZhT and Narodnoye khozyaystvo SSSR for relevant years. Daily locomotive kilometres exclude distances run assisting and light, and ancillary work.

Table 2.D   CHANGES IN RAIL TONNAGES BY AREA (AARC)

| | AARC % 1950-1975 | | | Region's share of traffic with East (%) | | | |
| | | | | Loadings | | Unloadings | |
	Loaded	Unloaded	Industrial prod.	1975	1969	1975	1967
North-West	+6.3	+6.0	+ 8.5	7.3	4.6	14.0	12.2
Centre	+4.4	+4.0	+ 8.0	11.0	9.1	19.6	16.3
Volga Vyatka	+5.1	+5.5	+ 9.4	x	x	x	x
C. Blackearth	+8.6	+6.8	+11.3	25.1	8.6	20.1	14.9
Volga	+7.4	+5.5	+11.1	x	x	x	x
N. Caucasus	+5.3	+6.2	+ 9.5	16.3	15.6	25.2	27.3
Urals	+4.8	+4.6	+ 8.7	x	x	x	x
W. Siberia	+6.3	+6.7	+ 9.6	x	x	x	x
E. Siberia	+7.5	+7.1	+10.4	x	x	x	x
Far East	+5.8	+6.1	+ 8.8	x	x	x	x
Don-Dnepr	+5.9	+5.8	+ 9.1	3.9	3.6	5.8	4.7
South-West	+6.3	+7.5	+11.0	4.2	2.8	7.6	7.3
South	+6.5	+8.5	+10.8	7.8	3.9	12.6	14.2
Baltic	+5.1	+7.1	+11.6	6.1	3.8	11.3	17.1
Trans-Caucasus	+6.8	+6.3	+ 8.3	6.8	5.4	10.6	11.5
Central Asia	+7.2	+7.5	+ 8.6	x	x	x	x
Kazakhstan	+8.2	+8.3	+10.5	x	x	x	x
Belorussia	+6.4	+7.5	+12.3	4.4	4.5	10.6	13.0
Moldavia	+7.5	+9.4	+12.5	11.2	10.0	8.8	10.5
USSR	+6.0	+6.0	+ 9.7				

Approximate long term increases :
  annual change in loaded tonnage = (annual change
  in ind.prod. x 3/4) - 2%
  annual change in unloaded tonnage = (annual change
  in ind.prod. - 4)%
The East is defined as the regions where rail
facilities are most under pressure, i.e. the 8
economic regions marked 'x'. In 1975, 7.3% of
tonnage loaded in the North-West went to these
regions, 14.0% of unloaded tonnage came from them.

Source : Narodnoye khozyaystvo for republics, and
Shafirkin (1978, pp.114-5)

GLOSSARY

ALH : average length of haul of freight (tkm/tons
    loaded)
Assisting : additional locomotive in the consist,
    because the train is too heavy for one
    locomotive
CTC : centralised train control, from a single point
    covering 100 km or more of route, which can
    control all signals/switches in its area, and
    monitor train movements
Circuity : distance travelled not allowed for in
    the tariff, e.g. to avoid a heavily used line;
    averages 2½% in the USSR. Tariff tkm + tkm in
    circuity = reported tkm
Consist : make up of a train; written or computer
    printed list of the freight cars in a train in
    correct sequence, together with their
    identification numbers and destinations
Cut : one or more freight cars from an inbound
    train, remaining coupled together because
    destined for a single classification track (and
    outbound train), and pushed over a marshalling
    yard hump by a switcher
Dynamic load : reported tkm divided by loaded car-km
    (cf. static load)
Exchange point : in 1975, about 125 places, normally
    yards, where two railways meet
Express freight train : train containing perishable
    and other priority freight, moving long
    distances with a minimum of marshalling
    (numbered 1001-1598)
EUROP : agreement between nine West European
    railways to use a large part of their freight
    car stock as if under 'common ownership'
Gamma factor : adapted from Sergeyev (1975, p.10),

equal to loaded car-km multiplied by tons loaded divided by reported tkm times freight cars loaded, or static load divided by dynamic load, averaging 1.1 to 1.2 for most large railways

Gross freight train weight : the sum of drivers' train reports giving net tonnage plus tare weight of the freight cars (excluding locomotives) times train distance, divided by freight train-kilometres

Inter-city passenger traffic : long distance passenger traffic (average journey in 1982 = 670 km)

LCL : 'less than car load' freight, under 10 tons, travelling at special tariffs, and requiring to be consolidated with other LCL at trans-shipment stations; in 1965, five eighths of LCL consignments and one sixth of LCL tonnage weighed less than 500 kg

Leg : distance between yards (plecho), calculated as total car-kilometres divided by number of yard operations

Light running : locomotives running without load for repositioning etc; in the USSR includes trains with less than 10 freight cars (Kozlov and Polikarpov 1981, p.217)

Load/unload : freight handling at terminals, whether railway owned and operated or at industrial sidings

Loadability : extent to which a given commodity utilises a freight car's tonnage capacity; coal has high, furniture low loadability

Loaded car trip : loaded car-km over a period divided by cars loaded

Loaded running % : where A = loaded car-km and B = empty-car km, calculated as A/(A+B); Soviet literature is not consistent on the definition of 'empty running %'

Local freight train : train linking local yards (about 750) and main yards (100 plus), numbered in series 3001-3398

Main yard : yard whose principal function is to handle trains to/from other main yards, with considerable mechanisation, high throughput, and special reporting requirements (form DO-24)

Marchroute : freight train avoiding remarshalling at at least one main yard, in various sub-types (despatched, staged - collecting or dropping off cars at various stations at either end of the trip, technical)

MBM : mineral building materials, the largest rail

freight category (about a quarter of total tonnage); about 60% is sand and ballast, cement over 10%, bricks 5%

Net freight train weight : as gross freight train weight, less tare weight of freight cars

OPW : agreement similar to EUROP (q.v.) for East European railways

Pick up freight train : trains moving between two yards collecting and leaving freight cars at way stations between those yards (<u>sborniye</u> : trains numbered 3401 -3498)

Pkm : passenger-kilometres (sum of passengers carried times individual distances)

Region : main operating division of a railway, 180 in 1977, with route lengths between 226 and 1675 km

Reload percentage : proportion of freight cars received which are reloaded at the unloading site without an intervening empty movement; in 1969 varying between 8.9% (Azerbaydzhan) and 21.8% (South Eastern)

Reported tkm : see 'circuity'

Reserve : difference between total stock and working park; for freight cars - cripples, held for special and seasonal traffic and uses (e.g. fire fighting, construction, housing, repair work); about 10% in the '60s, but thought to be increasing during the late '70s to at least 20%

RIV : 1922 agreement (Regolamento Internationale Veicoli) controlling the exchange of freight cars, and the tariff therefor

S & T : signalling and telecommunication(s) division, staff

Static load : tonnage loaded divided by cars loaded

Tariff tkm : tariff tonnage times tariff distance

Technical speed : speed of freight trains excluding time spent at way stations (derived from drivers' train reports)

Through freight train : trains moving between main yards and also marchroute trains (numbered 2001-2998)

Tkm : ton-kilometres (sum of tons loaded times distance)

Traffic : pkm and/or tkm

Train report : basic form recording gross and net tons, time in transit, distance travelled etc; in 1980 freight train reports numbered about 55 thousand per day

Transfer freight train : one which moves freight cars between yards in a node (like Moscow), numbered 3601-3798; in Russian, <u>peredatochniye</u>

Soviet Railways

Trip freight train : one which links a yard with a few subsidiary way stations (vyvozniye, numbered 3501-98)
UIC : International Union of Railways
Way station : station located between two yards with limited trackage for train formation and sorting; on single-track routes, trains can pass each other here
Working park: total stock less reserve
Yard-to-yard speed : technical speed including time spent at way stations (from driver's train reports)

REFERENCES

Abramov, A.P. (1974), Zatraty zheleznkykh dorog i tsena perevozki
Belen'kiy, M.N. (1974), Ekonomika passazhirskikh perevozok
Chudov, A.S. (1976), Sebestoimost' zheleznodorozhnykh perevozok
Danilov, S.K. (1977), Ekonomicheskaya geografiya transporta SSSR
Dewdney, J.C. (1976), The USSR, Dawson, Folkestone
Dmitriyev, V.A. (1980), Narodnokhozyaystvennaya effektivnost' elektrifikatsii zheleznykh dorog i primeneniya teplovoznoy tyagi
Gundobin, N.A. (1971), Spravochnik ekspluatatsionnika
Hunter, H. (1968), Soviet Transport Experience, The Brookings Institution, Washington
Hunter, H. (1983), The Soviet Railroad Situation, Wharton Econometric Forecasting Associations
Izosimov, A.V. (1967), Ulusheniye ispol'zovaniya osnovnykh fondov zheleznodorozhnogo transporta
Khandkarov, Yu. S. (1977), Vychislitel'nyy tstentr zheleznoi dorogi
Kozlov, T.I., and Polikarpov, A.A. (1981), Statistika zheleznodorozhnogo transporta
Kovrigin, A.G. (1978), Finansy zheleznodorozhnogo transporta
Mulyukin, F.P. (1975), Ekonomika zheleznodorozhnogo transporta
Povorozhenko, V.V. (1974), Expluatatsiya zheleznykh dorog
Sergeyev, E.S. (1978), Planirovaniye sistemy pokazateley ispol'zovaniya vagonnogo parka
Shafirkin, B.I. (1978, 1971), Ekonomicheskiy spravochnik zheleznodorozhnika
Silayev, (1975), Khozyaistvennyy raschet na zheleznodorozhnom transporte

Sotnikov, E.A. (1974), 'Optimization of prospective planning of classification yard development', Rail International, May 1974, 359-68

Sotnikov, E.A. (1980), 'Ways of intensifying classification yard operation', Rail International, July-August 1980, 429-47

Tikhonchuk, Yu. N. (1972), Ratsionalnoye raspredeleniye gruzovykh perevozok mezhdu zheleznodorozhnym i avtomobil'nym transportom

Tsarev, R.M. et al (1981), Nauchnye osnovy upravleniya, ASUZhT

Williams E.W. (1962), Freight Transportation in the Soviet Union

Perevozki gruzov, various authors (1972)

Problemy razvitiya SSSR, various authors (1981)

Transport i svyaz' (1972)

All Russian books published in Moscow, imprint 'Transport".

Chapter 3

THE TRANSPORT OF FUEL IN THE SOVIET UNION

David Wilson

Growth and Importance of Fuel Transport

Fuel is the most important commodity carried by the Soviet transport system. In 1982, 2,560 million tons of fuel (crude and processed) were carried an average 1,403 km giving a turnover of 3,586.5 billion ton-km - just over half the 7,107 billion ton-km carried by rail, road, river and sea transport and oil and gas pipelines.

The share of fuel in total transport turnover is growing, from just under 40% in 1970 to over 50% by 1982. So, while total transport turnover has been growing at an average annual rate of 5.0% since 1970, that of fuel has been growing by 7.1% a year (table 3.1).

Table 3.1 Transportation of fuel, billion ton-km

Year	All Transport*	Transport of fuel	Fuel as share of total, %
1970	3,961	1,582	39.9
1975	5,481	2,354	43.0
1980	6,781	3,375	49.8
1981	7,019	3,469	49.4
1982	7,107	3,587	50.5

* Rail, road, sea, river and oil and gas pipelines

A second reason, after tonnage, for the growing significance of fuel is the dramatic increase in the average length of journey. Since 1970, it has risen at an average annual rate of 3.2% from 965 to 1,403 km in 1982, with a particularly rapid growth of 4.3% a year during the second half of the 1970s. Since

90

1980 it has slowed to 1.5% a year.

This increase in the average length of journey of a ton of fuel is connected with the shift in the centre of gravity of the oil and gas industries to Tyumen' oblast in West Siberia, from the Urals-Volga region by the oil sector and from the Ukraine and Central Asia in the case of gas. In 1970, West Siberia accounted for only 4.7% of the USSR's gas (9.3 from 197.9 billion cu.m.) and 9.8% of its oil (31.4 from 353.0 million tons), but by 1983 the respective shares had risen to 50.3% and 59.9%. The nearest market for Siberian gas is the Urals region, over 1,000 km distant from the main producing gasfields, and most gas travels much further, through a system of pipelines running as far as the Centre region (e.g. Moscow), the North-West, the Baltic republics, Belorussia and the Ukraine. Nearly 48 billion cu.m./year of Siberian gas travels 4,450 km to the Western border of the USSR for export to the rest of Europe. A significant share of Siberian oil is piped to the refining complexes of the Volga region - Ufa, Kuybyshev, Saratov, Volgograd, etc. - although over 100 million tons a year goes through the Druzhba and Surgut - Polotsk pipelines to the refineries of the Baltic and Belorussian regions and to Ventspils for export. Between 1970 and 1982, the average distance travelled by oil (by all means of transport) rose from 989 to 1,642 km (i.e. by 4.3% a year) while that of gas increased from 611 to 1,404 km (7.2% a year).

Natural Gas Pipelines. Natural gas is now being pumped out of West Siberia through 13 inter-continental trunk pipelines in five corridors. These are:
(1) The Siyaniye Severa corridor consisting of three pipelines from Medvezhye and Urengoy deposits running south-westwards to the Peregrebnoye compressor station on the River Ob'. From here, they strike westwards across the northern part of the Ural mountains, down through the Komi ASSR (where they join the corridor's first pipeline, originating at Vuktyl) and on through Vologda, Yaroslavl', Kalinin and Smolensk oblasts to Belorussia, terminating in the western Ukraine at Dolina. Important branch lines run from Gryazovets to Moscow and Leningrad, from Torzhok to Leningrad, from Minsk to Gomel' and from Ivatsevichi to the Baltic republics. The corridor's capacity is 91 billion cu.m./year.

The Transport of Fuel in the Soviet Union

(2) The Tyumen' - Moscow corridor parts company with Siyaniye Severa at Peregrebnoye and proceeds southwards to Nizhnyaya Tura where it splits, with two pipelines continuing their southerly course through Sverdlovsk to Chelyabinsk and the other two following a westerly path through Perm', Kazan' and Gor'kiy to Moscow. The second Perm' - Kazan' - Gor'kiy line was completed in 1979.
(3) The Tyumen' - eastern Ukraine corridor branches from the Tyumen' - Moscow system at Nizhnyaya Tura and its two pipelines run through Kungur (Perm' oblast), Tuymazy (Bashkir ASSR) and Tol'yatti (Kuybyshev oblast) to Petrovsk, 100 km south of Penza. From here, it continues on a south-westerly course, terminating at Novopskov in Voroshilovgrad oblast. It is designed to satisfy the gas requirements of the eastern Ukraine, where the local output is slowly declining.
(4) The Tyumen' - south Urals - Volga corridor follows a completely different course out of West Siberia. It has two pipelines running south from Urengoy, picking up gas from Vyngapur on the way, and crossing the River Ob' at the Priobskaya compressor station 40 km east of Surgut. It then travels through Tobol'sk to Chelyabinsk before turning westwards to Dyurtyuli. Here, it joins the Tyumen' - eastern Ukraine corridor, which it accompanies to Petrovsk.
(5) The latest corridor was opened up with the completion of the Urengoy - Uzhgorod export pipeline (known in the West as the 'Yamal' or 'Siberian' pipeline) from West Siberia to the Czechoslovak border. The corridor follows a new path through Tyumen' oblast, avoiding Peregrebnoye and crossing the Ob' at Oktyabr'skiy, 75 km to the south. Joining the first three corridors at the Komsomol'skoye compressor station, it accompanies them as far as Nizhnyaya Tura, follows the Tyumen' - eastern Ukraine system to Kungur, then blazes a new trail through Pomary (Mariy ASSR), the Chuvash and Mordov ASSRs and Algasovo (Tambov oblast) where it intersects the Central Asia - Moscow corridor. It continues through Yelets and Kursk in the Central Blackearth region, and then follows a new path through Sumy to the Bar compressor station. Here, it joins the Soyuz pipeline from Orenburg and accompanies it to Uzhgorod on the western border of the USSR. By June 1984, a second pipeline had been laid along this corridor to Yelets, with most of its gas destined for Khar'kov oblast, and a third (also to Yelets) was more than one third completed. A

fourth, running from Yamburg rather than Urengoy was about to be started, and a fifth will also run from Yamburg to Uzhgorod to supply Eastern Europe with gas; this will be built in 1986.

Central Asia is the second most important gas-producing region after West Siberia, although its output has stabilised since 1975 at between 100 and 105 billion cu.m./year. Two corridors lead out of the region; one runs from Bukhara to the Urals with two pipelines and the other (the Central Asia - Centre system) takes four pipelines to Aleksandrov Gay in Saratov oblast (where one line peels off to Novopskov), Petrovsk, Algasovo, Ryazan' and Moscow. One of the lines originates at Bukhara, two at Mary where they tap the prolific eastern Turkmen fields, and the fourth collects and prepares casinghead gas from the western Turkmen and Mangyshlak oilfields before joining the other three at the Beineu compressor station.

The recent discovery and commissioning of two super-giant gasfields in the Turkmen republic, Dauletabad and Sovetabad, will allow the production of gas in Central Asia to grow substantially in the late 1980s. Accordingly, a new string of the Central Asia - Centre system was built in 1984, from Khiva to Aleksandrov Gay.

The production of gas by the Uzbek republic is likely to grow as the capacity of the Mubarek treatment plant, which cleans sulphur from the sour gas of the Mubarek field, is expanded. Most of this additional gas will flow eastwards to Tashkent and along the northern edge of the Tien Shan mountains to Frunze and Alma Ata. The construction of the necessary pipeline was started in 1984.

The only other area planned to have a big increase in gas production is Astrakhan oblast, where production from the first stage of 6 billion cu.m./year will begin in 1986. The gas will flow south to Mozdok in the North Caucasus in an existing pipeline which formerly carried Caspian gas in the opposite direction.

With gas production expected by the official Long Term Energy Programme to continue growing strongly to the end of the century before it stabilises, probably at about 1,000 billion cu.m./year, a further 12-14 trunk lines out of Siberia can be expected in addition to those mentioned above. Perhaps two of these will run eastwards to supply the Kuzbass conurbation, which is currently served only by a line bringing a small amount of casinghead gas from the Middle Ob' oilfields. The large cities of Omsk, Novosibirsk, Krasnoyarsk and Irkutsk should also be served by a pipeline, the construction of which may take place during the next plan period 1986 - 90. Otherwise, all the new pipelines will flow westwards.

Between 1970 and 1983, the length of the Soviet trunk gas pipeline network grew from 67,500 to 155,000 km at an average rate of 6.6% a year. All the pipelines out of West Siberia have diameters of 1,420 mm and are each able to carry 33 billion cu.m./year. The total length of 1,420 mm diameter pipelines has grown from nothing in 1970 to 3,600 km in 1975 and more than 12,000 km in 1981[1]. The eleventh five-year plan target of 48,000 km of gas pipelines [2] includes the construction of seven 1,420 mm diameter pipelines from Urengoy to the western USSR of total length 25,000 km. It is certain that this target will be met, with pipelines now being built at rates unprecedented in world practice, and five of the seven lines were already completed by June 1984. Although the five-year plan initially predicted that only six of the seven pipelines would be completed during the plan period, it now seems possible that the seventh (i.e. Yamburg - Yelets) will become operational. In 1983, 11,000 km were built, and the 1984 plan[3] of 10,500 km, or 12,200 km including gas liquids pipelines[4], is likely to be overfulfilled.

Crude Petroleum Pipelines. There has been an equally dramatic increase in the length of the oil pipeline network since the development of the West Siberian oilfields began. In 1967, the first pipeline carrying Tyumen' oil was laid from Ust'-Balyk to the Omsk refinery, and in 1970 the eastwards flow along the existing Trans-Siberian pipeline was reversed. Since 1970, oil pipelines from the Middle Ob' oilfields have followed four corridors :
(1) Aleksandrovskoye - Tomsk - Anzhero-Sudzhensk has one pipeline, which was completed in 1972. In 1973, it was extended to Krasnoyarsk as part of a plan to build a pipeline carrying Tyumen' crude to the Pacific seaboard for Japan, although this plan was subsequently dropped. With the start-up of the Achinsk refinery in 1982, the pipeline's capacity usage rate was increased by 6 million tons/year and it has now been extended to Irkutsk[5]. Part of its oil serves the extension of the Angarsk refinery, and the rest is trans-shipped to rail for haulage to

the Khabarovsk refinery; a new trans-shipment complex opened for this purpose at Irkutsk in August 1983[6]. Thus the Trans-Siberian pipeline now has two strings carrying Tyumen' oil.

(2) West Siberia - Volga. Two pipelines have been built from the Nizhnevartovsk field, which consists mainly of the Samotlor deposit, to the Volga region. They both have a diameter of 1,220 mm and can carry up to 90 million tons/year when working at maximum capacity. The first, completed in 1973, runs to Al'met'yevsk in the Tatar ASSR, from whence the oil is distributed through a local network of smaller pipelines to the refineries in the Bashkir ASSR and Kuybyshev oblast. The second pipeline runs to Kuybyshev and was completed in 1976. Its oil is subsequently distributed throughout the western USSR, through lines running to Novorossiysk and Tuapse on the Black Sea (completed 1974), to Lisichansk, Kremenchug and Odessa (1977), from Lisichansk through Tikhoretsk to Groznyy, and finally from Groznyy to Baku in a pipeline completed in 1983[7]. The Nizhnevartovsk - Kuybyshev - Lisichansk - Groznyy - Baku line is over 4,000 km long and constitutes the USSR's longest oil pipeline[8].

(3) West Siberia - Central Asia. The Ust'-Balyk-Omsk pipeline was extended to the Pavlodar refinery in Kazakhstan in 1977, and the Pavlodar - Chimkent stage of 1,642 km was completed in 1983. Although the Chimkent refinery has not yet become operational, oil is being pumped to Chimkent where it is trans-shipped to rail and sent to the 3 million tons/year Fergana refinery. The small Alty-Aryk refinery is also being supplied in this way[9]. The pipeline is to be extended further to Chardzhou, but there is no evidence of construction work having been started, and when the Chardzhou refinery becomes operational in 1985 it will probably be supplied initially by rail.

(4) West Siberia - western USSR. The 3,390 km Surgut - Polotsk pipeline with an eventual capacity of 90 million tons/year was completed in February 1981[10]. Two months later, the new refinery at Mazeikiai (Lithuania) was receiving oil through the Polotsk - Mazeikiai extension[11]. The Surgut - Polotsk pipeline supplies the Perm', Gor'kiy, Yaroslavl', Moscow, Ryazan' and Novopolotsk refineries as well as Mazeikiai and also supplies oil for export through the Ventspils terminal.

In 1982, the construction began of the second pipeline to follow the corridor. Nearly 2,500 km

The Transport of Fuel in the Soviet Union

long, it will supply the Perm', Gor'kiy, Yaroslavl'
and Moscow refineries, and will terminate at Klin,
80 km north west of Moscow. It will originate at the
new Kholmogorsk oilfield, 200 km north-east of
Surgut, and although the start-up date was
originally put as 1984, the annual plan for 1984
foresaw its commissioning only as far as
Nizhnekamsk[12]. Completion of this pipeline will
permit oil production in West Siberia to exceed 500
million tons/year by 1990.

This is the only major crude oil pipeline under
construction at the present time, because oil
production is planned to grow very slowly from 603
million tons in 1980, and 616.3 in 1983, to 630
million tons in 1985. Other new lines, such as
Kenkiyak - Orsk, are designed to connect new
oilfields to the existing pipeline network. The
emphasis is now on building oil product pipelines;
of the 16,000 km of oil pipelines planned to be
built over 1981-85 [13] (compared with 13,100 km
completed during the previous five years), some
12,000 km are to be product pipelines[14]. This is
five times the length of product pipelines laid
during 1976-80, when only 1600 km actually
became operational. However the annual plan targets
cast doubt on whether the five-year plan target is
still attainable; the 1983 plan, for example, was to
build only 1,400 km of product pipelines[15].

Petroleum Product Pipelines. The most important
product pipelines to be completed during the current
plan period are:
(1) Gor'kiy - Novki - Ryazan' - Tula - Orel - Kromy.
The first stage of 228 km from Gor'kiy to Novki was
completed in 1981[16], and the Novki - Kromy stage
(650 km) was finished in 1984 [17].
(2) Petropavlovsk - Kokchetav - Tselinograd. This
500-km pipeline[18] is a branch of the existing Ufa
- Novosibirsk product line. It will carry gasoline
and diesel, refined from Bashkir and Tyumen' crude at
the Ufa complex of refineries, mainly to the farms
of northern Kazakhstan. Its construction began in
1982[19], and the first stage of 180 km from
Petropavlovsk to Kokchetav was commissioned in May
1983[20].
(3) Nikol'skoye - Voronezh, 202 km, was completed in
April 1983. It will deliver gasoline and diesel from
crude refined at Ufa and Kuybyshev to the farms of
the Central Blackearth region[21].
(4) Groznyy - Budennovsk, 183 km, was completed

in late 1984. It will carry low octane gasoline to the Prikumsk petrochemical plant at Budennovsk, thereby saving R 100,000 per year compared with the cost of rail transport[22].

(5) Lisichansk - Donetsk - Zhdanov (280 km)[23]. The Lisichansk - Donetsk section came on-line in early 1984[24] and, together with its branch lines, is 170 km long. It also supplies Gorlovka, Kramatorsk and Krasnoarmeysk and carries 1½ million tons of products a year, saving 29,000 rail car movements. The Donetsk - Zhdanov stretch is now being built, and is due for completion in late 1984. The line will then be extended to Melitopol' and the Crimea[25] for a further 425 km.

(6) Sineglazovo - Sverdlovsk, 215 km, which feeds products to the largest city in the Urals.

(7) Travniki - Kustanay - Amankaragay, 375 km, is to be built during 1984 and 1985, and will deliver gasoline and diesel to the farms of northern Kazakhstan.

(8) The gas liquids pipeline from Komsomol'sk on the West Siberian gasfield to Minibayevo in the Tatar ASSR also appears to rank as an oil product pipeline according to the five-year-plan. It will carry products from the West Siberian casinghead gas refineries, principally the 8 billion cu.m./year plant at Nizhnevartovsk, to the Volga region for further processing[26]. It will be 1,451 km long when completed.

However, most of the 12,000 km planned for 1981-5 will consist of a great many fairly short pipelines carrying products from oil product bases administered by Glavneftesnab to large customers such as power stations, chemical plants, airports etc. A typical example is the 60 km line built in 1983 to serve the Kiev area with gasoline and diesel[27].

Since 1970, the crude and product pipeline network has grown from 37,400 km to 72,900 in 1982 at an average annual rate of 5.7%. Growth was most rapid during 1971-5 when pipelines were completed at a rate of 3,840 km/year; this compared with 2,620 km a year over 1976-80 and 1,600 km/year in 1981-2. The slowdown in construction rates is attributable partly to the very slow rate of increase in oil production, and partly to the consequential emphasis on building gas pipelines.

Pipelines are by far the best way of transporting oil and gas. For gas, they are the only way, with liquefaction prior to rail transportation not being an economical or safe option. For crude

The Transport of Fuel in the Soviet Union

oil, pipelines have a number of obvious advantages over rail transport:
(1)   They are cheaper, with the cost of piping oil up to 12 times cheaper than rail when large diameter pipelines are employed.
(2) They are less labour intensive, especially in the loading and unloading processes.
(3) They provide a continuous flow rather than a series of discrete flows to the refinery, cutting its requirements for storage tanks and facilitating its operation.
(4)   Losses in transit are greatly reduced.

Table 3.2   Soviet oil pipelines

	1970	1975	1980	1981	1982
Length, thousand km	37.4	56.9	69.7	70.8	72.9
Crude	30.7	46.6	57.8		
Products	6.7	10.3	11.9		
Volume, million tons	339.9	497.5	630.2	637.9	645.0
Crude	314.6	458.0	574.0		
Products	25.3	39.5	56.2		
Turnover, billion					
ton-km	281.7	665.8	1,196.8	1,263.2	1,306.8
Crude	259.8	638.8	1,160.4		
Products	21.9	27.0	36.4		
Productivity, tons					
input per km of					
line	9,088	8,743	9,042	9,010	8,848
Crude	10,247	9,828	9,930		
Products	3,776	3,835	4,722		
Average length of					
journey, km	828	1,338	1,899	1,980	2,026
Crude	826	1,395	2,021		
Products	865	683	647		

Sources: Narodnoye khozyaystvo SSSR 1982, p.307
Problemy razvitiya transporta SSSR, table 7.1

Rail Borne Fuel.   The Soviet railways have quite enough problems without having to handle ever-increasing volumes of fuel that could be shipped more easily by other methods. Consequently, it has been official policy for a number of years
98

now to take into account the need to relieve railway congestion when planning the structural and geographical fuel balance of the USSR. This can be achieved by:

(1) Transferring the shipment of crude oil and its products to pipelines.

(2) Building gas pipelines to those regions of the USSR which import coal over particularly long distances; the Baltic republics and Belorussia, which have been burning Kuzbass coal railed over 4,000 km, are now receiving rapidly growing quantities of gas through the Siyaniye Severa system of pipelines from West Siberia.

(3) Constructing nuclear power stations primarily in those regions at greatest distance from the sources of possible increases in fuel production such as the Central, Central Blackearth and Ukrainian regions.

(4) Developing large power complexes on the basis of coal, gas and hydro resources in the eastern regions and transmitting the power westwards.

(5) Converting to gas those facilities which burn coal or fuel oil carried long distances by rail.

In 1975, 59 million tons of crude oil (12% of output) was moved by rail with traffic of 113.6 billion ton-km, giving an average journey of 1,925 km; 330.6 million tons of products and 367.8 billion ton-km were carried an average of 1,113 km. By 1980, the tonnage of crude had fallen by 39.3% to 35.8 million tons [28], and it declined further to 33.5 in 1981[29]. It should now be less than 27 million tons/year, and this sharp decline is due basically to the construction of three more pipelines. The Samgori - Batumi line is now carrying nearly 3 million tons a year 430 km from Georgia's only oilfield to Batumi. It was completed in 1980 with oil reaching the Batumi refinery in November of that year[30]. The Tyumen' - Yurgamysh line carries 5 million tons/year of oil which originates from the Shaim field in West Siberia. This was formerly delivered to Tyumen' by a pipeline built in 1966 and then railed for 1,800 km to the Volgograd refinery[31]. The new pipeline extends the Shaim-Tyumen' line for 250 km as far as the Trans-Siberian pipeline at Yurgamysh, and it began operating in 1981[32]. In 1982 the construction of a branch line from the Kuybyshev-Tikhoretsk pipeline (which runs through Volgograd), and the reconstruction of the Kuzmichi pumping station, allowed Shaim oil to be pumped directly to Volgograd. The completion of the Groznyy - Baku

The Transport of Fuel in the Soviet Union

pipeline in 1983 also allowed up to 6 million
tons/year of crude to be taken off the railways.
    Other recently built crude oil pipelines, such
as Perm' - Al'met'yevsk in 1982[33], are designed
principally to improve the manoeuvrability of oil
flows through the network while others, such as
Kaspiysk - Orsk[34], are intended to connect up new
deposits.

Table 3.3 The transport of fuel by rail

	1970	1975	1980	1981	1982
Volume (million tons)	950.0	1,141.1	1,188.6	1,175.7	1,183.8
Oil and oil					
products	302.8	389.0	422.7	428.8	425.1
Coal	613.9	716.9	731.6	713.7	725.3
Coke	33.3	35.2	34.3	33.2	33.4
Traffic (billion					
ton-km)	802.0	1,008.8	1,089.6	1,073.4	1,066.2
Oil and oil					
products	353.9	481.4	460.8	456.9	450.6
Coal	424.6	498.0	598.8	587.2	587.1
Coke	23.5	29.4	30.0	29.3	28.5
Average journey (km)	844	884	917	913	901
Oil and oil					
products	1,169	1,238	1,090	1,066	1,060
Coal	692	695	818	823	809
Coke	706	835	875	882	853
Share of fuel in					
total rail trans-					
port, %					
Tonnage	32.8	31.5	31.9	31.3	31.8
Ton-kilometres	32.1	31.2	31.7	30.6	30.8

Source : Narodnoye khozyaystvo SSSR 1982 (Firewood,
peat, shale, etc. excluded above; tars, waxes, etc.
included: in 1975, 6% and 1.9% of total fuel tkm)

    The transfer of oil products from rail to
pipeline transport is most economical for light and
middle products such as gasoline, naphtha, jet fuel
and diesel. The pipelining of residual fuel oil
(39.1% of 1975 petroleum products rail traffic) is
more difficult because of its higher viscosity,
which also means that a greater pumping capacity is
needed to move it along the pipeline. However, it

still pays to pipe fuel oil because transit losses are much less than on the railways, since tank cars have to be steam-cleaned after each journey, with a residue lost in the process. In fact, it is well known that consignees leave much more than this residue of fuel oil in tank cars, especially when it has congealed and is difficult to discharge. In this case, steaming stations have to stop work until the offending tank car has been taken off the track, and its contents removed and poured into an oil product lake. All steaming stations have these lakes; when they are full, their contents are handed back to the Goskomnefteprodukt organisation, which administers the distribution of oil products. Plan targets have even been set for this procedure, and in 1982 the target of 200,000 tons was slightly underfulfilled, with only 192,000 tons collected. Often, the product lakes are burned to prevent them overflowing into rivers.

The Chernikovka-Vostochnyy oil loading depot[35] near Ufa has the USSR's largest station for washing and cleaning tank cars. It suffered a raid in early 1983 by members of the People's Control Agency to discover why its annual throughput of tank cars had declined from 300,000 several years ago to 159,600 in 1982. On the day the inspectors arrived, the station was supposed to wash 1,000 wagons; work went smoothly until three from the Crimea were found to have some 80 tons of fuel oil left in them, although their despatch ticket declared them to be empty. The fuel had congealed, and the Crimean workers, held responsible for delays in the despatch of wagons, had simply declared them empty and sent them on their way. Work on the line stopped to remove the wagons, and before long the other two lines were stopped for similar reasons - the entire station had ground to a halt. Wagons were found to contain 7 tons of gasoline from Serebryanoprudsk oil depot near Moscow, 18 tons of higher olefins from Sten'kino, and large amounts of engine oil and chemicals; a tank car from Baladzhary even contained 3 tons of treacle. The loss to the country is much greater than the value of the wasted fuel. While wagons containing fuel remnants are being dealt with, there are shortages of railway rolling stock and shutdowns of plant at refineries because the railways are unable to shift the products. During 1982, Ishimbay refinery (Bashkir ASSR) with a rated capacity of 12 million tons/year came to a complete halt on two occasions, and at other refineries, shutdown was only averted by

The Transport of Fuel in the Soviet Union

'freeing' gasoline storage tanks by pouring their contents into tanks containing fuel oil.

Natural Gas Distribution Strategies. The pattern of distribution of West Siberian gas throughout the USSR is partly determined by the need to reduce long-distance shipments of coal. The major industrial areas nearest to the West Siberian gasfields are the Kuzbass and the Urals, and given the considerable cost of building large-diameter trunk pipelines, it may appear more economical to fully supply these regions with gas before supplying the most distant areas such as Belorussia and the Baltic republics. In fact the Kuzbass is served only by the small casinghead gas pipeline from Samotlor, and the Urals region is not particularly well served; much of the gas flowing through the Urals and Volga down the various pipeline corridors continues on to Moscow or the Ukraine. Most of the big power stations in the Urals and Volga regions are coal or oil-fired - Reftinsk (3,800 MW), Troitsk (2,500 MW) and Yuzhnoural'sk (1,000 MW) burn coal, Zainsk (2,400 MW), Iriklinskiy (1,800 MW), Karmanovo (1,800 MW) and the Brezhnev Heat and Power Plant (1,000 MW) use fuel oil, and Verkhniy Tagil (1,625 MW) burns both coal and oil. The only large power station in these two important industrial regions fired with gas before the current conversion programme began was Sredneural'sk (1,200 MW). By contrast, most of the power stations in the Central region burn gas, usually in a dual-fired regime with oil.

All the gas distribution scenarios considered by planners at one time or another have made Central Russia and the western republics the first priority. When the Siberian fields were first developed, it was foreseen that they would be yielding 280 billion cu.m./year by 1980, of which 50 billion cu.m. would be burned in the Urals and 230 billion would pass down the Siyaniye Severa corridor to central Russia. Although this plan was amended, it is likely that by 1983, when Siberian production reached 270 billion cu.m., the share-out was similar.

Gas consumption in the westernmost parts of the USSR is already high by national standards and is continuing to rise. In the Baltic republics and Belorussia, 4.5 million dwellings are now served with gas; this is 150% more than the 1.8 million of 1970 and accounts for 85% of all dwellings in these republics compared with 77% for the USSR as a whole. Yet more branch lines from the Siyaniye Severa system are still being built, and in Latvia alone, gas consumption is planned to grow from 1.7 billion cu.m. in 1981 to 3.2 in 1985, i.e. an increase of 88% compared with 38-39% for the USSR as a whole over the same period. In Lithuania, gas consumption is planned to rise by 50% over 1981-85 [36] to 5 billion cu.m./year[37], permitting the import of coal from other regions to be reduced by at least a million tons. The programme of connecting all towns and .settlements to the gas supply network was completed in September 1983. Belorussia consumed 5.36 billion cu.m. in 1982, compared with 4.75 in 1980 and 2.86 in 1970, and consumption is planned to rise by 130% during 'the next few years'[38], which is thought to mean the period 1984-90. In Leningrad oblast, gas accounts for 60% of fuel consumption, with supplies increasing by 2.4 billion cu.m./year during 1976-80. This saved 40,000 rail wagons a year of coal and oil imports from other regions[39].

Pipeline construction in the western republics is continuing at a frenetic pace. The completion has recently been announced of the lines from Kaunus to Kaliningrad (270 km), from the Siyaniye Severa line to Mogilev (90 km)[40], from Riga to Yurmala and Riga to Panevèžys, and from Minsk to Gomel' (307 km). The construction of the Riga - Daugavpils pipeline (200 km) with a projected capacity of 1 billion cu.m./year has begun [41] with completion planned for 1985, and the second string of the pipeline from Izborsk (Pskov oblast) to the Incukalns underground reservoir (219 km) started up towards the end of 1984.

Additional supplies of gas will permit further reductions in the supply of Kuzbass coal which has to be railed over 4,000 km to the western republics. While the Donbass is much closer, its hopes of raising output to 223 million tons in 1985 are unlikely to be met, with production dropping below 200 million tons in 1982 and still falling. The planners have tried (unsuccessfully) to keep the Donbass power stations fully supplied with steam coal by halting deliveries beyond the Ukraine. In any case, Donbass coal is becoming much more expensive to produce, as the mines go deeper and the seams become thinner. Kuzbass coal is competitive in the western USSR, even after shipment over immense distances, because of its very low production costs. Natural gas is cheaper still, which is why the

western republics are continuing to receive a high
priority for new distribution pipelines. Since 1980,
many boilers and heat-and-power plants have been
converted to gas from coal and this trend will
continue until coal is forced entirely out of the
fuel-energy balance.

Electricity

Nuclear Power. The siting of nuclear power
stations is determined with similar motives in mind.
All the 63,940 MW capacity at nuclear stations under
construction at the present time, or being extended,
will be installed in the European part of the USSR,
including 22,000 MW in the Ukraine (where 5,880 are
already available at Chernobyl', Rovno and
Konstantinovsk) and 15,000 MW in the Central Region
(2,000 so far at Smolensk and Kalinin). A further
4,500 MW is to be commissioned at Ignalina in
Lithuania (where the first set of 1,500 is now
generating) and 14,000 MW in the Volga region.
Nuclear stations can be located practically anywhere
with an adequate water supply, and the stations
operating at the present time have been sited
specifically where the availability of primary fuels
is most limited, such as Leningrad, Ignalina,
Kalinin, etc.
During the 1981-85 plan period, 24-25,000 MW
of nuclear capacity is planned to come on-stream. By
mid-1984, only 9,380 had been commissioned and the
five-year plan is now infeasible. This also puts at
risk the forecast that 80,000 MW will be generating
480 billion KWh, or 27% of predicted electricity
output, in 1990. Capacity installed since the
beginning of 1981 consists of: the fourth set of
1,000 at Leningrad and the third of 440 at Kola
(both in the North West region), the third and
fourth of 1,000 MW each at Chernobyl' (Ukraine), the
first and second of 440 MW at Rovno (Ukraine), the
first 1,000 at Smolensk and the first (1,000) at
Kalinin (both in the Central region), the first
(1,000) at the south Ukraine station in
Konstantinovka, the third of 1,000 MW at Kursk
(Central Blackearth region) and the first of 1,500
at Ignalina (Lithuania). Sets which were due for
start-up in 1984 included Kola-4 (440 MW),
Smolensk-2 (1,000), South Ukraine-2 (1,000),
Balakovo-1 (1,000) in the Volga region, and
Zaporozh'ye-1 (1,000 MW) in the Ukraine.
All the sets completed or planned for the

1981-5 period have two aims - firstly to relieve the chronic electricity shortages which have bedevilled the USSR for many years now, and secondly to permit the closure of small coal-burning power stations. The only exception is Balakovo, located in the rapidly growing Volga region where there are practically no coal-fired facilities to close. The relative importance of these motives varies from station to station. At one extreme, the Kursk station will devote practically all its additional power to the KMA territorial production complex centred on the Oskol direct reduction steel plant and its associated industries, for which electricity demand will grow very rapidly. At the other extreme, Ignalina serves an area of the USSR which is growing relatively slowly and with comparatively little heavy industry, for which electricity demand is not particularly high. Although some of its power may be exported to Eastern Europe, most of it will replace electricity from small coal-burning plant. This will also imply a greater electrification of the economy with electricity replacing other forms of heat in many industrial processes and dwellings, thereby permitting many coal-fired boilers to close. It is significant that the Long Term Energy Programme adopted in the spring of 1984 foresees a greater electricity intensity accompanying a lower fuel intensity of the economy in the years to come. In the western republics, facilities using coal rather than oil will be singled out for scrapping, because the coal is brought in over vast distances; moreover, there is an excess availability of fuel oil from the Mazeikiai, Novopolotsk and Mozyr' export-oriented refineries.

For other nuclear stations, the relative significance of the coal-saving and increased-output motives is not so clear cut. In the Ukraine, the motive of replacing coal is not so much prompted by the need to obviate long rail hauls as by the poor performance of the Donbass coalfield since the late 1970s (particularly the steam coal sector); and the realisation many years ago that coal-mining in the Donbass is not the best way to utilise scarce labour resources, when such huge reserves of a cheap, clean and easily-used fuel like gas are available. Consequently the coalfield is being starved of capital investment funds while an increasing share of the Ukraine's electricity will come from nuclear stations. Chernobyl' is to be expanded from 4,000 to 6,000 MW, Rovno from 880 to 2,880, South Ukraine from 1,000 to 4,000, and new stations are under construction at Zaporozh'ye (4,000), Khmel'nitskiy (4,000) and the Crimea (4,000). The 4,000 MW Rostov station in the North Caucasus may also supply the Ukraine.

Other Sources of Electricity.   As well as building nuclear stations at the point of demand, transmitting electricity from the source of energy supply is another option. Much project and costing work has been carried out in Soviet research institutes with the aim of comparing the options, and there is no clear evidence that one is more economic than the other. Consequently a compromise has been reached, with both options being pursued. There are four sources in the resource-rich eastern zone of the USSR at which power can be generated for transmission westwards – the Ekibastuz and Kansk-Achinsk coalfields, the West Siberian gasfield and the East Siberian hydro-electric stations.

The Ekibastuz field in Kazakhstan currently produces 72 million tons/year of coal which is classified by the Soviets as hard coal, but which would be regarded as brown coal in the West. It has a heat rate of only 4,250 kcal/kg, but is strip-mined using very large bucket-wheel excavators, giving very low production costs. It is therefore economical to transport the coal as far as the Urals, and 64 million tons of coal were railed out of the coalfield in 1983 to about 20 power stations in the Urals, West Siberia, and other parts of Kazakhstan. However, the future of the field is based on the eventual construction of 20,000 MW of mineside generating capacity in five stations of 4,000 MW each.   At the moment, the first station is nearing completion with seven of its eight sets of 500 MW each connected. Three more stations are in varying stages of construction, and the fifth is to be sited at some distance from the existing pits near Lake Balkhash.

The power from the first station is distributed through 500 KV lines to Omsk (to the north-west), Ust'-Kamenogorsk (south-east) and through Karaganda and Agadyr' to Dzhezkazgan (south-west). Another 500 KV line is planned to run from Agadyr' to Alma Ata and connect with the isolated Central Asian grid.

At the same time, three super-high-tension lines are under construction. The highest priority is being given to a 1,150 KV AC line which will run for 1,200 km before joining the Urals grid near Magnitogorsk. The first section of 450 km to

Kokchetav is now energised[42] and the next stage to Kustanay (400 km)[43] will start up in 1985. The whole line should be completed by the end of 1986 and it will carry 33 billion KWh/year, i.e. the whole output from Ekibastuz-2 and from the first two sets of Ekibastuz-3 power stations. Simultaneously, another 1,150 KV AC line is being built over 600 km to Barnaul in West Siberia[44].

A 1,500 KV DC transmission line is to follow that going to the Urals, but it will terminate at Tambov, covering a distance of 2,500 km. Little has been heard about it recently although work is still said to be taking place[45]. However, the Long Term Programme refers only to a giant 'ring' of 1,150 KV lines gathering power from Ekibastuz, Kansk-Achinsk, the Siberian gasfields and the Angara-Yenisey hydro stations. Power for the Far East will be siphoned off, and two 1,150 KV lines will carry electricity to the western USSR. The ring is to be completed in the early 1990s, but just how the 1,500 KV DC line fits into the long-term plan is not clear.

The other large fuel and energy complex based on coal is KATEK on the Kansk-Achinsk coalfield. The plans for KATEK have varied considerably over time, and initially some 10 power stations of 6,400 MW each were to be built before the end of the century. They would burn 360 million tons/year of Kansk-Achinsk coal, which cannot be economically transported because of its low heat rate and tendency to spontaneous combustion. The power would hopefully be despatched westwards through lines of up to 2,250 KV. At the moment, the only firm plan is to build one station at Berezovo towards the western end of the field, burning coal from the nearby Berezovo opencast mine. Although the mine is now producing 2 million tons/year of coal acquired during the overburden removal operation, the power station's first set is unlikely to be commissioned before the end of 1985. Meanwhile, research work continues at a feverish pace on the creation of transmission lines, insulators and sub-station equipment of sufficiently high capacity to make the full project feasible.

There are also plans to build more power stations based on West Siberian gas. At present, one station has been completed at Surgut with a capacity of 3,360 MW. It is intended to serve the Middle Ob' oilfields, with some help from the Reftinsk station (Urals) which transmits power along a 500 KV line through Tyumen' and Tobol'sk to Surgut. Power also reaches the oilfields through a 220 KV line from the

Kuzbass through Tomsk to Nizhnevartovsk. The Surgut-1 power station also supplies the more northerly gasfields through a 500 KV line to Novyy Urengoy, from which a 220 KV extension is now being erected to the Yamburg gas deposit.

However, this is barely sufficient to cover the needs of the oil and gas fields where vast amounts of electricity are needed to operate drilling rigs, pumping stations for waterflood systems, and oil and gas treatment plants etc. Power cuts are endemic, and each of the 16 sets at Surgut-1 station has been installed behind rather than ahead of the growth in demand for power, which now exceeds 80 million KWh/day at peak periods. This situation should improve in 1985 with the commissioning of the first 800 MW set at the new Surgut No. 2 station which, like No. 1, will burn casinghead gas from local oilfields. By 1990 it will be completed with 6 sets of 800 MW each, making 4,800, and another power station of similar size at Nizhnevartovsk will also be operating. Finally, a 4,000 MW station designed to burn natural gas is under construction at Tikhyy near Urengoy. This should be completed by 1990, and during the first half of the 1990s, the total generating capacity of power stations in Tyumen' oblast should reach nearly 20,000 MW.

West Siberian oil production is likely to stabilise at 500 million tons/year in 1990, while gas output should grow strongly to 550 billion cu.m. in 1990 before flattening out at about 800 billion cu.m./year by the end of the century - these are the implications of the Long Term Energy Programme of 1980-2000. There will thus be a big local surplus of electricity, even assuming no more power stations are built, and this will be tapped by the 1,150 KV ring mentioned earlier, for transmission to the European USSR.

Finally, firm decisions are now being taken to build massive hydro-electric stations on the rivers of East Siberia, principally the Yenisey and its tributaries. The large stations now operating - Krasnoyarsk (6,000 KW), Bratsk (4,500), Ust'-Ilimsk (3,840) and Sayano-Shushenskoye (now 3,840, but likely to reach 6,400 MW by the end of 1985) serve the needs of East Siberia and the Far East, but the new stations will be specifically intended to transport power to the European USSR. Firm plans have so far been accepted for Turukhansk, Sredneyeniseysk and Boguchany. Turukhansk will be built on the Nizhnyaya Tunguska. It will have 20 sets of 1,000 MW each[46] and the first of these is tentatively planned to begin generating by 1995[47]. Some 46 billion KWh/year[48] will be despatched westwards. Construction will begin by the end of 1984. Sredneyeniseysk will be on the Yenisey near its confluence with the Angara. Part of the 31 billion KWh/year[49] from the 7,500 MW station will be used locally in a territorial production complex devoted to timber processing, but perhaps 25 billion KWh/year will be available for transmission westwards starting from 1995. Boguchany (4,000 MW) is located on the Angara, and construction has already started; it will begin producing power in 1990.

According to our forecasts[50], nuclear power stations will account for 65% of the 220,000 MW of new generating capacity likely to come on-line between 1980 and 2000, and hydro stations for a further 17%; over 144,000 MW of nuclear and 38,000 MW of hydro capacity will be installed. If these nuclear and hydro policy options are not pursued and thermal stations of similar total capacity are built, mostly in the western zone of the USSR, and burning fuel transported from the eastern zone, then an additional 330 million tons of coal equivalent of fuel must be produced annually by the end of the century. The additional transport burden would amount to 800-850 billion tkm, over 11% of the total USSR freight traffic by all modes in 1982. Though much of this burden would fall on the gas pipelines, a significant share would be claimed by the rail haulage of coal and fuel oil.

The Power Station Conversion Programme. While considerable reductions in the potential transport burden will be made with the development of nuclear and hydro power stations and the Ekibastuz, Kansk-Achinsk and West Siberian fuel and energy complexes, the immediate emphasis is on converting those power stations and boilers which burn fuel oil and coal to natural gas. This should bring some relief to the railways, who are mainly responsible for transporting fuel oil and coal although the waterways also play an important role; in 1969, a rare year with complete statistics, the split of traffic in coal/coke and petroleum products (i.e. excluding crude oil) was rail - 88%, inland waterways - 6%, coastal shipping - 3%, and pipelines - 3%. It should be remembered, however, that relieving the railways of some of their work is not the principal motive behind the current conversion programme.

When planners were drawing up the eleventh five-year plan, they had a number of variants of the fuel mix to consider. One variant foresaw oil production continuing to rise, from 603 million tons in 1980 to perhaps 700 in 1985, with output in West Siberia increasing from 312 to 460 million tons, i.e. by nearly 30 million tons a year. This would involve a massive programme of infrastructural development in sparsely populated areas, in the Tarko Sale region for example, because the additional oil output would have to come from a large number of scattered small to medium sized deposits. The construction ministries charged with the provision of housing, amenities and roads etc., would face a truly daunting task on a scale which they were not prepared to contemplate. Consequently, the variant finally adopted planned for a much slower rate of growth of oil production to 630 million tons in 1985 of which 399 would come from West Siberia at an average annual increase of 17.4 million tons over 1981-85. At the same time, there would be an extraordinary surge in annual gas production from 435 to 630 billion cu.m. with West Siberian output rising from 156 to 357 billion cu.m., or by 40 billion/year. This scenario was much more acceptable to the construction ministries because all the increase in gas production would come from just one supergiant deposit - Urengoy.

The most important consequence of the current five-year plan for oil is that export levels can only be maintained by restricting the growth of domestic demand. This is to be achieved partly by fuel-saving, but primarily by reducing the amount of fuel oil burned in power stations from 150 million tons in 1980 to 120 in 1985 and further, to 75 million tons in 1990 (according to the Long Term Energy Programme). The affected stations will be converted to natural gas, and the fuel oil thus saved will be subjected to secondary refining processes so as to squeeze greater volumes of gasoline and diesel from it. Hence a steadily rising demand for automotive fuels, which has been increasing by over 5% a year, will be met from a more slowly growing production of crude oil.

The programme of converting oil-burning facilities to gas has been under way for a number of years now. Initially it concentrated on raising the share of gas in the fuel mix of dual-fired (i.e. fuel oil and gas) power stations. Dual-fired stations are built so as to make use of gas during the summer months when demand for it declines (the high cost of gas pipelines gives a big incentive to reduce demand fluctuations), and in the past an oil/gas power station set would normally burn oil for seven months and gas for five. There is evidence that in some stations this ratio has been reversed; for a 300 MW set working for 6,000 hours a year, this provides a saving of nearly 100,000 tons of fuel oil a year.

However, the role of dual-fired power stations as equalisers of gas demand will be less important in the future, because of the expansion in the number of underground gas reservoirs; these accumulate gas during the summer and release it in the winter. Consequently, power stations can be converted to operate entirely on natural gas, and in some areas this process has already begun. In 1983, condensing stations and combined heat-and-power plants (CHPPs) in Khar'kov, Perm' and Yaroslavl' were converted. At Khar'kov, the large CHPP No. 5 was connected by a 150 km branch line to the Urengoy-Uzhgorod pipeline[51] and this eliminated the annual rail haulage of 400,000 tons of fuel oil from the Lisichansk refinery, over 225 km away.

The plans for 1984 are more ambitious. The 1,800 MW Karmanovo station in the Bashkir ASSR is undergoing conversion to gas, which will be brought through a 140 km branch line from Birok on the Urengoy-Chelyabinsk-Petrovsk trunk line. It used 2.6 million tons of fuel oil a year[52], 1 million of which came from the obsolete Karmanovo refinery (which has now been scrapped) and 1.6 million tons by rail from the Ufa refineries over a distance of 270 km. The 420 MW Pechora power station in the Komi ASSR has been converted to casinghead gas from the Usinsk oil deposit; previously it burned 1.5 million tons/year of fuel oil, which was railed over 1,500 km from Yaroslavl'. This was a clear case of conversion designed to assist the railways rather than save fuel oil, and was particularly urgent because the capacity of the Pechora station is to be doubled shortly. The Leningrad Yuzhnyy CHPP (500 MW) has also been converted to gas. It is not clear whether its fuel oil was railed from Kirishi refinery or whether it travelled through the Kirishi - Krasnyy Bor product pipeline.

While it is a major aspect of economic policy to reduce the consumption of fuel oil, the saving of coal has also acquired greater importance since 1980. This is because of the failure of the coal industry to get anywhere near its 1985 target of 775

111

million tons[53] - in 1983 production declined by 2 million tons to 716 million. The resulting shortage is particularly acute in the Donbass, where the operation of some power station sets has been suspended, and power cuts have seriously affected the region. Some stations have only been kept going with coal railed 3,900 km from the Kuzbass[54]. It is ironic that over 50 billion cu.m./year of gas is produced in the eastern Ukraine and four large-diameter gas pipelines from Urengoy terminate in, or pass through, the region. So, several coal-burning stations are now being converted to gas. The most important of these is Uglegorsk station[55] at Svetlodar. It has a capacity of 3,600 MW and burns 7.2 million tons (coal equivalent)/year of coal and oil. A 48 km branch line has been laid to bring Siberian gas to the station [56] and the first sets have now been converted. The Zuevka No. 2 station was originally intended to burn anthracite waste, but there have evidently been technical problems because, although only two of the projected 8 sets of 300 MW each have been installed, the station is already being converted to natural gas. Industrial and municipal boilers are also being converted to gas in the Donbass at places like Snezhnoye[57].

Changes in Flows of Rail Borne Coal. The fact that coal-fired stations like Uglegorsk, located in the heart of the coalfield, are now being converted to gas rather than the most distant customers for Donbass coal, indicates that planners are more concerned with achieving the maximum possible volume of conversion, rather than reducing the boundary of provision of Donbass coal. This might suggest that the relief of rail congestion by taking coal loads off the railways does not carry a particularly high priority when compared with other economic policies; or perhaps the boundary of provision is shrinking anyway as a result of the sheer non-availability of supplies, rather than from conscious decisions by the planners.

The Urals is a region where the import of coal is unavoidable. It is a traditional coal producing area with five coalfields, all of which are now in a state of decline. Although it is relatively close to the West Siberian gasfields, it burns a mere 11.9 billion cu.m./year[58] of gas, less than 10% of its total fuel consumption, and the volume of coal railed in from the Kuzbass, Karaganda and Ekibastuz is growing rapidly, leading to congestion on the railways. The USSR's largest fossil-fuel condensing power station at Reftinsk burns 12 million tons/year of Ekibastuz coal and some 3,000 rail wagons a day[59] (carrying up to 200,000 tons/day) flow down the Pavlodar - Tselinograd railway en route for the power stations of the Urals. In the past, the service has been erratic with many wagons often missing or in the wrong place, and trains stranded between stations, resulting in large quantities of coal not reaching their destinations on time.

Three policies have been adopted to improve things. Firstly, the adoption of a new system of labour organisation at Ekibastuz - the so-called 'complex shifts' system - has halved the idling of wagons awaiting loading, thereby increasing the number of wagons loaded by 10,000 during the first five months of 1984.

Secondly, the creation of direct delivery express trains (i.e. trains comprising wagons all intended for the same destination) has obviated the break-up and re-formation of trains at intermediate yards. At Tselinograd, a special despatcher service has been set up to ensure that direct express trains can be formed in both directions, and these now account for practically the entire movement of coal on the Pavlodar - Tselinograd - Magnitogorsk section. The 225 km line between Tselinograd and Atbasar used to be a major bottleneck with trains taking 12 hours to cover it; now they move four times as fast, averaging 75 km/hour. The big increase in the average speed of trains during 1981-83 contributed towards the 40% increase in haulage on the Tselinnaya railway during this period.

Thirdly, super-heavy trains have been formed experimentally, with the maximum load growing from 10,000 tons a few years ago to a record 33,220 tons recently. The record train was 4.5 km long and its 320 wagons carried almost the entire daily requirement of the Reftinsk power station. Train loads of 16,000 tons now appear to be sent most days, and super-heavy trains carried 3 million tons in the first quarter of 1984, saving the work of 500 train crews and cutting electricity consumption by 5 million KWh.

While the share of fuel in total transport traffic will grow significantly in the future from the 50.5% of 1982, it can be anticipated that the share of fuel in rail traffic will decline from the 30.8% of 1982. This is because the volume of coal/coke and oil carried by rail is not likely to rise by much over the 1,184 million tons of 1982, and could well fall; the trend has been static since 1977.

Our forecasts for the coal industry[60] predicted output rising to 815 million tons in 1990 and 1,000 million tons in 2000, when 60 per cent would be produced in opencast mines and 40 per cent would come from deep mines: these expectations were recently confirmed[61]. Much of the strip-mined coal will be burned on-site, including almost all that from the Kansk-Achinsk field and 70 million tons/year from the eventual 170 million produced at Ekibastuz.

The most important factor determining the future burden of coal upon the railways is the development of slurry pipelines. Firm plans for coal pipelines in the USSR currently consist of a 225 km pipeline to be built from the Inskoy mine near Belovo in the Kuzbass to the Combined Heat and Power Plant No. 5 in Novosibirsk[62]. It will be similar in design to the USSR's only operating slurry line, which runs 10 km from the Yubileynaya mine near Novokuznetsk (also in the Kuzbass) to the West Siberian Metallurgical Plant. The cost of building this line, and that of the plant for crushing the coal and mixing it with water, was recouped in less than two years. Transportation of coal by pipeline is much cheaper than by rail, avoids the time-consuming processes of loading and unloading, and avoids spillages in transit.

In December 1982, the relevant committee of Gosplan approved the project for the Inskoy-Novosibirsk line. Construction of a minehead plant began in mid-1982; some 3 million tons/year of coal will be crushed there, and this will then be mixed with an equal quantity of water and pumped along the pipeline. This will be built of 1,420 mm diameter laminar pipe, manufactured at the Vyksa pipe plant. The power station, with a capacity of 1,200 MW, will produce 6 billion KWh/year and will be supplied with all its fuel from the pipeline.

The committee also studied a proposal for a system of pipelines stretching for more than 4,000 km from the Kansk-Achinsk field to the Moscow area. At the moment, they foresee the line carrying up to 60 million tons/year of coal. It may be slurried with a mixture of water and methanol which enables the volume of coal to be increased. Methanol can be removed from the coal fairly easily, and the pipeline would therefore allow appreciable quantities of methanol to be pumped westwards for petrochemical feedstocks, or for addition to petrol.

Changes in Flows of Rail Borne Oil. For oil, an absolute decline in the volume and turnover of rail transportation is foreseen, despite a probable increase in crude oil production from 616.3 million tons in 1983 to 660 in 1990 and 700 million in 2000[63]. Our forecast of a very gradual increase in output until the next century was confirmed by the official Long Term Energy Programme. The haulage of crude oil by rail will virtually cease, especially when oil consumption in the Far East (estimated at 15 million tons/year in 1980[64] and 18 in 1983) rises to the point where a large expansion of the Khabarovsk refinery and the laying of a crude oil pipeline from Angarsk to Khabarovsk becomes economically feasible. The movement of oil products by rail is more difficult to transfer to pipelines because many flows are simply too small (see Table 3.2) to make the construction of a pipeline worthwhile, especially if there is an existing rail track. Soviet planners are probably expecting that 200-250 million tons/year of oil (including 5-10 million tons of crude) will travel by rail in 2000 compared with 425 million in 1982. With many of the long-distance product flows transferred to pipelines, it is unlikely that the average length of haul by rail will exceed 500 km (1,060 km in 1982) giving traffic of 100-125 billion ton-km (450.6 in 1982). In the short term progress will be slow, with the volume planned to decline by no more than 10 million tons in 1985 compared to 1980, i.e. from 422.7 to at least 412.7 million tons[65].

The projected decline in the rail transport of oil products will be assisted by a dramatic change in the product mix of refineries, with the output of fuel oil cut to 150 million tons/year (under 24% of the anticipated refinery throughput in 2000 compared with 42.6% in 1980); the production of diesel will rise to 230 million tons (36.5% compared with 25.0%), and that of gasoline to 130 million tons (20.6% compared with 15.0%). It is much easier to pump gasoline and diesel through pipelines than to pump fuel oil becaue they travel faster and can be batched more easily. Fuel oil is more likely to be sent through specialised pipelines connecting oil product bases with CHPPs and large municipal boilers - very few power stations will be burning fuel oil by the end of the century.

The Transport of Fuel in the Soviet Union

If, against expectations, the West Siberian oilfields prove to be disappointing, then oil pipeline traffic will grow extremely rapidly with a major push eastwards by oilmen. This is not foreseen by the Long Term Energy Programme, which confirms that West Siberia will be the main source of oil until the 21st century. Exploratory work has been stepped up recently in East Siberia, mainly in the Lena-Anabar Trough and the Nepski Svod, where oil has been discovered, but it looks increasingly likely that the Pri-Caspian Depression to the north-east of the Caspian Sea will attract more attention during the next decade. Production has already started at the giant Tengiz deposit, and huge oil reserves are forecast at depths of 5,000 to 7,000 metres. If the technology to develop these sub-salt deposits can be created, then the Pri-Caspian fields of Aktyubinsk and Ural'sk oblasts will prove very attractive indeed, given their proximity (850 km) to the USSR's most important refining regions of the Urals and Volga.

In conclusion, it seems inescapable that the oil, gas (and possibly coal) pipeline networks will expand very rapidly during the next 15 years, enabling the burden of fuel transport on the railways to decline sharply.

REFERENCES

[1] Pravda 9 January 1982
[2] Izvestia 18 November 1981
[3] Ekonomicheskaya gazeta 12, 1984
[4] Ekonomicheskaya gazeta 3, 1984
[5] Pravda 15 November 1982
[6] Pravda 12 August 1983
[7] Pravda 10 June 1983
[8] Izvestia 31 December 1982
[9] Izvestia 15 March 1983
[10] Pravda 23 February 1981
[11] Izvestia 28 April 1981
[12] Ekonomicheskaya gazeta 3, 1984
[13] Izvestia 18 November 1981
[14] Pravda 28 February 1981
[15] Pravda 2 January 1983
[16] Pravda 3 September 1981
[17] Ekonomicheskaya gazeta 3, 1984
[18] Pravda 21 May 1982
[19] Pravda 22 November 1982
[20] Pravda 8 May 1983
[21] Ekonomicheskaya gazeta 18, 1983

[22] Pravda 22 April 1983
[23] Ekonomicheskaya gazeta 3, 1984
[24] Ekonomicheskaya gazeta 6, 1984
[25] Izvestia 9 February 1984
[26] Izvestia 14 February 1984
[27] Izvestia 24 January 1983
[28] From Problemy razvitiya transporta SSSR table 7.5
[29] Economist Intelligence Unit (EIU) Quarterly Energy Review: Soviet Union and Eastern Europe (QER) 2, 1982, p.31
[30] Pravda 21 November 1980
[31] Pravda 28 February 1982
[32] Pravda 8 November 1981
[33] Izvestia 19 December 1982
[34] Pravda 30 December 1982
[35] EIU QER 2, 1983, p.40
[36] Pravda 20 September 1983
[37] Vilnius radio 3 September 1982
[38] Izvestia 11 March 1984
[39] Izvestia 5 March 1982
[40] Moscow radio 9 January 1982
[41] Tass 2 April 1984
[42] Ekonomicheskaya gazeta 14, 1984
[43] Ekonomicheskaya gazeta 29, 1983
[44] Pravda 5 January 1984
[45] Izvestia 8 December 1983 and Izvestia 12 August 1983
[46] Leningrad radio 1 February 1984
[47] Ekonomicheskaya gazeta 23, 1984
[48] Moscow radio 4 November 1982
[49] Tass 16 March 1983
[50] Wilson, D.C. (1983), Demand for Energy in the Soviet Union, Croom Helm, London, p.84
[51] Ekonomicheskaya gazeta 3, 1984. This source says that the pipeline is only 35 km long, which is obviously wrong
[52] Pravda 5 January 1984
[53] Izvestia 15 November 1981
[54] Izvestia 8 April 1984
[55] Izvestia 13 January 1984
[56] Izvestia 15 November 1983
[57] Izvestia 25 January 1984
[58] Izvestia 19 September 1982
[59] Pravda 2 June 1984
[60] Wilson (1983), p.259
[61] EIU QER 2, 1984, p.18
[62] EIU QER 1, 1983, p.24
[63] Wilson, (1983), p.269
[64] Wilson, D.C. (1980), Soviet Oil and Gas to 1990, Economist Intelligence Unit, London, p.24·

The Transport of Fuel in the Soviet Union

[65] EIU QER 2, 1982, p.31

- " -

Problemy razvitiya transporta SSSR (1981),   various
     authors, Transport, Moscow

Chapter 4

SOVIET SHIPPING : ITS IMPACT ON THE WEST

Paul E. Lydolph*

The Problem as Perceived by the West
Since 1961 the Soviet commercial fleet has increased
its deadweight tonnage by 358 per cent, which has
raised its ranking among world fleets to seventh in
deadweight tons   and third in number of ships
(Tables 4.1 and 4.2). Consequently, the presence of
Soviet ships has become much more noticeable along
the major sea lanes of the world, in many of which
they are now engaging heavily in cross-trading in
order to run their ships efficiently and gain hard
currency. About   7870 vessels belonging to the
Soviet fleet of Morflot now make more than 25,000
calls at 1200 ports in 124 countries every year.
     Aside from the political consequences that
might imply, the economic impact of a powerful new
competitor, operating under entirely different rules
from those long-established by shipping conferences,
has brought chaos into many conferences and
bankruptcies to some lines. The situation has been
particularly destructive for British lines which had
controlled almost 15 per cent of the world's
shipping capacity in 1961 when the British flag flew
over by far the largest fleet in the world. Since
then the fleet flying the flag of the United Kingdom
has dwindled to only 4.8 per cent of the world total
tonnage, sixth place among fleets (Table 4.1).

The Soviet Response
The Soviets claim that the Western world is itself
responsible for the shipping crisis. They say that
the Soviet decision to develop a major merchant
fleet stemmed from the refusal by many Western
shipping lines immediately following the Cuban
missile crisis in 1962 to handle Soviet exports and
imports that were growing at a rate of about 30 per

119

Soviet Shipping : Its Impact on the West

Table 4.1   Merchant fleets of the world totalling more than 9,000,000 deadweight tons. Oceangoing ships of 1000 gross tons and over as of January 1, 1983.[1].

Country of Registry	No.of Ships	Deadweight Tons[2]	Rank Dead-weight	Rank Ships [3]
World	25482	671,093,000	--	--
Liberia	2145	140,293,000	1	4
Greece	2604	68,868,000	2	2
Japan	1755	63,665,000	3	5
Panama	3141	56,288,000	4	1
Norway	577	36,237,000	5	10
United Kingdom	816	32,067,000	6	6
USSR	2482	22,457,000	7	3
USA (Private)[4]	573	21,647,000	8	11
France	318	17,422,000	9	18
Italy	605	15,747,000	10	8
Singapore	588	12,042,000	11	9
Spain	517	11,942,000	12	12
China (PRC)	811	11,798,000	13	7
Germany (FRG)	439	10,381,000	14	15
India	385	9,826,000	15	16
Korea (ROK)	474	9,552,000	16	13
Brazil	346	9,176,000	17	19
Saudi Arabia	192	9,048,000	18	--

[1] Excludes ships operating on the Great Lakes of North America and inland waterways of the world as well as special ships, etc., and merchant ships owned by any military force.

[2] One metric ton = 2205 pounds. Sources vary. According to the USSR   Register in 1982 sixteen USSR   marine shipping companies operated 2000 vessels with a total deadweight tonnage of 19,512,000

[3] In numbers of ships, the Netherlands ranks 14 and Cyprus 17. Many other countries rank ahead of Saudi Arabia, whose tonnage consists primarily of a relatively few large oil tankers

[4] Government owned ships numbered 259 with a deadweight tonnage of 2,756,000

Source : Compiled from data supplied by the U.S. Department of Transportation, Maritime Administration, Office of Trade Studies and Subsidy Contracts, Division of Statistics.

cent per year. Determined not to be vulnerable to
such actions again, the Soviets initiated a massive
buildup of their merchant marine through
construction in their own shipyards and purchases
from other countries such as Poland, East Germany,
and Finland.

Thus, they built their fleet to serve their own
export-import needs. But to operate these fleets
efficiently, they sought cargoes in other countries
to fill ships on return voyages. Eventually they
engaged in some cross-trading when excess capacities
were available, but in all cases, they point out,
their cross-trading activities, as well as their
shipping in general, has been insignificant compared
to some other leading fleets in the world. They
point out that their merchant fleet is carrying
approximately 220 million tons of cargo per year
which is about 7 per cent of the world total, and of
this the freight lifted in cross-trades makes up
less than 1 per cent of the world seaborne trade.
Thus, the Soviets contend that the present crisis in
world shipping has been brought on by the rapid
expansion of fleets much larger than their own
coupled with a stagnation or decline in world
trade.

Table 4.2 Percentage changes in deadweight
tonnages of active merchant fleets totalling more
than 20,000,000 tons, ranked according to
deadweight as at 1 January, 1983. Only oceangoing
ships of 1000 gross tons and over are included.

Country of	Percentage Change			
Registry	1961-76	1976-81	1981-83	1961-83
World	224	18	+ 2.5	290
Liberia	663	16	- 8.5	706
Greece	415	85	- 1.0	844
Japan	572	- 2	+ 2.7	578
Panama	262	72	+48.2	823
Norway	197	-19	- 6.2	125
U.K.	118	-23	-24.1	27
U.S.S.R.	271	20	+ 3.2	358
U.S.A.	6	41	+ 2.4	53

Source : Compiled from data supplied by the
United States Maritime Administration.

Soviet Shipping : Its Impact on the West

Growth of Fleets and World Trade
One cannot deny that practically all the shipping
companies of the world began a rapid expansion of
merchant fleets in the early 1960s in anticipation
of expanding world trade which never materialised.
While the Soviet fleet grew by 358 per cent between
1961 and 1983, the much larger fleets of Greece and
Japan grew by 844 per cent and 578 per cent
respectively (Table 4.2). Open registry expanded
tremendously, now accounting for approximately 30
per cent of world tonnage. During this period, the
deadweight tonnage flying the convenience flag of
Panama alone expanded by 823 per cent and that
flying the flag of Liberia by 706 per cent. Flags of
convenience largely explain the poor showing of
USA flag ships in Table 4.2. Many of the ships
flying the flags of convenience are owned by
American companies or by multinational corporations,
large shares of which are owned by American
interests. According to the US Maritime
Administration, in 1982 481 American vessels
totalling 50.7 million deadweight tons carried other
flags. The Soviets say that this makes the American
fleet total at least 80 million deadweight tons and
point out that figures compiled by UNCTAD (United
Nations Commission on Trade and Development) are
much higher, which makes the USA the world's largest
owner of merchant fleets. The Soviets argue that
much of the fleet flying the flag of Liberia belongs
to owners in the USA, Hong Kong, Greece, and some
countries in Western Europe. They point out that the
Maritime Administration of Liberia controls the
fleet registered under its flag from a computer
centre in Reston, Virginia, and that ships owned by
American operators and registered under all flags of
convenience are controlled from New York City.
    It is difficult to ascertain exactly who owns
what since so many ships are not flying the flags
of ownership and so many ships are owned by multinational groups. Also ownership may not determine
who is operating the ships, since so many crews are
hired from Third World countries such as Bangladesh
and the Philippines where wages are low and
attention to such things as safety and insurance is
minimal.
    World trade began to stagnate and decline
during the first half of the 1970s, particularly
after the oil embargo of 1973, during which time
world shipments of liquid cargo shrank by 147
million tons (7.1 per cent) and dry cargoes
decreased by 86 million tons. During the rest of the

1970s seaborne trade fluctuated up and down year by year, but since 1979 it has declined consistently, which is without precedent. The fall of 8 per cent in 1982 was also without precedent. Since 1979 total world trade has decreased by approximately 500 million tons, a drop of around 30 per cent. At the same time the continuing shift in world trade patterns towards shorter routes has caused the ton-mile index to decrease even more. This has been particularly true in the demand for tanker capacity, which has declined for the fifth successive year. Also the deteriorating state of the world's steel-making industry has had a devastating effect on the level of iron ore shipments, particularly from South America. During 1982 alone world ore movements decreased by about 10 per cent and coal shipments by about 4 per cent.

Excessive dry cargo capacity has resulted in a decline of time-charter rates from a high of about $19.00 per deadweight ton per month in May 1980, to about $6.50 in February 1983, for general cargo ships. Bulk carrier rates run much less, some of the large bulkers less than half as much. Single voyage tanker rates of 15-25,000 ton capacity on the Mediterranean-UK route have declined from more than 400 points in October 1979, to about 140 points in February 1982[1]. The freight rates for 'handy-sized dirty' tankers in 1982 were less than one-third of late 1979 figures. By the end of February 1983, approximately 90 million deadweight tons were laid up, about 22.5 per cent of the world's ships. Obviously, world shipping is in the greatest crisis ever known. With this economic backdrop, it is easy to understand why even a small interjection by a new competitor is met with indignant outcry, particularly if that competitor is viewed as using unfair advantages.

As can be seen in Table 4.2, most of the growth of the larger fleets of the world took place during 1961-76, and since then has tapered off or even declined. Although orders for new-buildings were greatly reduced immediately after the 1973 crisis, ships inbuilding continued to be turned out in record numbers until about 1976, and in some cases considerably later than that. Among the home-owned fleets, only that of Greece continued to expand rapidly during 1976-81, and part of that apparent gain might have been due to shift of registry from flags of convenience to the national flag. During the last two years even Greece has shown an absolute decline. The USA showed a belated spurt during

Table 4.3 Sizes of fleets totalling more than 20,000,000 d.w.t., by category, as at 1 January 1983, and deadweight tonnage changes, 1981–83

Country of Registry	Freighters			Bulk Carriers			Tankers		
	No.	Deadweight 1000 tons	Per cent change	No.	Deadweight 1000 tons	Per cent change	No.	Deadweight 1000 tons	Per cent change
World	14280	124994	+ 3.1	5215	208153	+12.4	5583	336142	− 3.0
Liberia	486	5713	− 3.4	854	43861	+ 5.8	797	90658	−14.4
Greece	1241	12676	−14.2	909	29240	− 3.0	407	26801	+ 9.4
Japan	689	6456	±0	506	23590	+ 6.8	552	33591	+ 1.0
Panama	2055	17085	+28.6	662	20580	+82.7	393	18436	+36.3
Norway	153	1685	−15.0	143	9952	− 3.8	261	24535	− 6.1
UK	321	3651	−22.9	164	8883	−19.8	322	19473	−25.9
USSR	1802	11261	+ 0.2	175	3604	+14.8	458	7480	+ 3.2
USA	247	4371	− 4.3	25	777	+33.3	293	16434	+ 3.8

Source : Compiled from data supplied by the United States Maritime Administration

1976-81.

Since 1976 there seems to have been a major shift of registry to the convenience flag of Panama, particularly of freighters. The Panama flag now flies over the largest freighter fleet in the world (Table 4.3). On the other hand, the convenience flag of Liberia has experienced an absolute decrease during the last two years. This might be due primarily to the adoption by the country of the international convention fixing minimum crew wages. Some shipowners might have transferred ships from Liberia to some other convenience flags such as Panama, Singapore, or other small countries such as Costa Rica, Honduras, the Bahamas, the Netherlands Antilles, or Maldive Islands, which are belatedly trying to cash in on revenues available from this source.

Some of the strongest competition to long-established fleets during the last few years has come from rapidly developing new fleets in some of the non-Western, non-Soviet bloc countries. Between 1981 and 1983, Saudi Arabia's fleet increased by 213 per cent, which suddenly shot the Arabian fleet into eighteenth place in the world. The Saudis are currently aggressively pursuing a large share of the shipping of project and container freight from various parts of the world to the Middle East. During the same two years the South Korean fleet increased by 52 per cent, that of Brazil by 19 per cent, the People's Republic of China by 16.5 per cent, and India by 6.5 per cent. The small Turkish fleet increased by 50 per cent in 1982 alone. Since its independence, Singapore, partially due to its acting as a flag of convenience, has become a major fleet, ranking eleventh in the world in 1983.

In contrast to some other leading fleets, the fleet of the USSR has expanded much in line with that of the total fleet of the world (Table 4.2). Most of the expansion took place prior to 1976, and since then has tapered off. At present the stated intent and apparent practice of the Soviets is to up-grade their fleet with replacements of new modern-type ships with only a slow net increase in tonnage. So far the Soviets have been hampered in competition for world freight by the continued use of small, inefficient vessels. Most of their freighters are less than one-third the size of those of leading nations, and most of them are break-bulk type. The Soviets are beginning to replace these with larger container, ro/ro, and LASH ships.

125

Soviet Shipping : Its Impact on the West

By the end of 1982 Morflot apparently owned 45 ro/ro ships and planned to add others that would total 5 per cent of the Soviet fleet by 1985. They owned 5 barge cariers built in Finland of the LASH type, and barge carriers were planned to amount to 2 per cent of the Morflot tonnage by 1985. Morflot owned approximately 125 full container ships. Container and ro/ro tonnage is planned to double during the present five-year plan[2]. Mitsui OSK Lines of Japan says that by 1985 the USSR will add 170 dry cargo ships, 50 container ships, 64 ro/ro ships, a large number of tankers, and atom powered LASH vessels. Morflot currently owns about 350 tankers and is adding new ones of 65,000 ton capacity, which means that they are relatively small. These are to be chartered out and may play an important role in carrying future supplies of Middle East oil to East European countries and Third World non-oil producing clients.

Much effort is being placed on the construction of ships and port facilities for the Northern Sea Route. The Soviets have said that the Northern Sea Route is the only possibility for supplying new northern developments, and estimate that by the year 2000 this route may handle as much freight as the Trans-Siberian railway. Nuclear ice breakers and reinforced freighters that can break through ice one metre thick are being put into operation. The first nuclear powered LASH vessel is under construction in Leningrad for service along the Northern Sea Route from Murmansk to the mouth of the Yenisey River and up the Yenisey to Dudinka and perhaps beyond. Servicing ports such as Dudinka several hundred kilometres up the Siberian rivers from the Arctic coast has always been a problem, because ocean-going ships used along the Northern Sea Route could not navigate the shallower and more ice-clogged waters of the rivers. Barge carriers seem to be the solution to this difficulty.

Western Objections to Soviet Practices

Fleet Composition and Cross-Trading. Western shipping lines note an ominous composition of the Soviet fleet that belies those statements to the effect that the fleet has been developed primarily to serve Soviet foreign trade. They point out that the freighter fleet comprises more than 10 per cent of the world general cargo capacity which is at least four times the Soviet foreign trade needs, while the bulk carrier and tanker fleets are inadequate. Dry and liquid bulk commodities account for 75 per cent of foreign trade cargoes, while the Soviet dry bulk and tanker fleets make up only 49.4 per cent of tonnage, clearly inadequate to handle this trade. The freighter fleet makes up the other 50.6 per cent, and in deadweight tons comprises the third largest freighter fleet in the world, after Panama and Greece. In comparison, 81.8 per cent of the total world fleet consists of bulk carriers and tankers, and only 18.9 per cent of freighters.

Western observers can only conclude that the Soviets have built their fleet primarily to serve cross-trade purposes to gain hard currency in the most lucrative trade routes of the world while leaving less remunerative cargo, including many of their own bulk exports and imports, to other flags, often on a time-chartered basis. It has been reported that in 1975 of a total of 58 Soviet lines 20 were involved almost wholly in international cross-trading[3]. During the 1970s they became particularly active in the US trade across the Atlantic and across the Pacific, in Europe-Far East trade, in Europe-East Africa trade, and in Far East-Australia trade. Most recently they have become quite active in Europe-Central America trade.

G. Smalley[4] estimates that currently Morflot is earning about 500 million dollars worth of hard currency per year by cross-trading dry bulk cargoes and another 480 million by cross-trading general cargo in addition to about 110 million dollars a year by the Trans-Siberian Land Bridge and 51 million dollars a year by cruise ships. This is in addition to about 400 million dollars a year by carrying their own imports and exports, for a grand total of almost two billion dollars per year. He points out that it would be very difficult for the Soviets to make up this two billion dollars with additional exports of raw materials, many of which would have to come from remote areas with high costs of exploitation. On the other hand, he points out that this aggregate sum derived from all modes of transport to foreign countries does not compare with the potential hard currency to be realized by the Siberian gas pipeline.

However, G. Maslov, President of the Soviet shipping company, Sovfracht, has stated that the amount earned by Soviet owners in liner and tramp cross-trades covers less than one-third of the expenditures in foreign currency for the carriage of Soviet exports and imports in foreign flag liner and tramp tonnage[5]. T. Guzhenko, Minister of the USSR Merchant Marine, has stated that in 1982 Soviet cross-trade liftings amounted to only 23.8 million tons, down from 36.6 million tons in 1974, and that non-Soviet ships, including cross-traders, in 1982 lifted more than 120 million tons, or 51 per cent of USSR exports-imports. Cross-trades made up 14.2 per cent of total Soviet liftings in 1981 and only 10.7 per cent in 1982, while Soviet foreign trade accounted for half of Soviet liftings and domestic coastal trade for more than one-third. It is significant that the Soviets always report their own cross-trading aggregating both tramp bulk commodities and liner and general cargoes, apparently in order to conceal their considerable activities in the higher priced general cargo trade. Similarly, they report foreign liftings of their own foreign trade aggregating both those liftings by trading partners and by cross-traders, apparently to obscure the small amount available to cross-traders.

No one faults the Soviets for chartering-out some of their bulk carriers. The Soviets commonly time-charter much more bulk carrier tonnage from foreign flags to carry bulky exports of oil and ores and imports of grain. However, it is a different story in the high-value general cargo liner trade, where the Soviets generally monopolize 80-90 per cent of their own foreign trade commodities while engaging heavily in cross-trading among the most lucrative general cargo liner trades of the developed countries of the world. Smalley estimates that Morflot cross-trades annually about 6 million tons of general cargo, which uses about one-third of their general cargo tonnage and almost all the tonnage of modern cellular-container and ro/ro ships[6].

Only recently have the Soviets slowed the expansion of their freighter fleet. Table 4.3 shows that during 1981-3 it grew by only 0.2 per cent, while the bulk carrier fleet grew by 14.8 per cent, and the tanker fleet by 3.2 per cent.

Aside from economic considerations, Western observers point to the utility of the Soviet fleet showing the flag in many ports of the world and generally to serve multipurpose political, military, and intelligence operations. Many observers point out that the unspecialized nature of most of the Soviet ships, and their relatively small size, make them easily convertible to military operations if necessary, many of them being adaptable to landings without port facilities. The fact that they are totally Soviet-owned and government-controlled would alleviate any problem of conversion from commercial to military craft. By contrast, most of the ships owned in the Western world, often by multinational companies, often manned by Third World crews, and often flying under registries foreign to their owners, could hardly be relied on for military operations in time of war.

There is little consensus of opinion among Western observers as to the primary purposes of the buildup of the Soviet fleet. Mitsui OSK of Japan contends that in order of importance they are: (1) to meet military considerations; (2) to influence developing countries; (3) to foster links with friendly trade partners; and (4) to earn foreign exchange. David Scrivener contends that the main Soviet concern is to achieve immediate domestic and foreign economic targets. In that context, the buildup of the fleet must compete with all the other economic concerns of the nation for scarce investment capital, and therefore the emergence of a comprehensive Soviet maritime policy directed towards attainment of command of the seas and ruthless undermining of Western economies will probably bog down in bureaucratic conflict. He says that neither present capabilities nor the thrust of naval doctrine appear to aim at wartime control of the world's major sea lanes[7]. Many specialists in the United States contend that the primary purpose of the Soviet commercial fleet is to alleviate hard currency deficits. Minister Guzhenko claims that the Soviet fleet plays an active part in the implementation of the peace programme outlined by the Soviet Communist Party.

Freight Rates. Soviet lines have operated primarily as outsiders, and in order to gain entry into major shipping lanes have had to cut freight rates considerably below those agreed upon by the conferences operating along the routes. In most cases rate reductions have been no more than 10-15 per cent, which is in line with tacitly agreed upon rate structures for other independents. But occasionally on certain highly competitive and lucrative routes the Soviets allegedly have cut rates as much as 50 per cent. The ultimate threat in such cases is that many long-established lines will be forced out of business, leaving the routes to the mercy of future actions by the USSR, which then may raise rates and slash services. In actuality, where conferences have engaged in rate wars, it appears that the Soviets do have some bottom line below which they will not go, and have consequently pulled out of such competitive routes, switching their ships to more remunerative areas.

On the other hand, the Soviets have often quoted higher than conference rates for bulky, low-value freight that would return less profit per carrier tonnage. General cargo shipping is 5-7 times as profitable per ton as bulk cargoes are; thus, conference lines have often found themselves with low-profit cargoes while Soviet independents skim the cream off the top.

In response to charges of rate cutting, the Soviets have stated that many Western lines offer rebates to shippers and forwarders at least as great as Soviet rate cuts. Investigations by the U.S. Federal Trade Commission have found some truth to these accusations.

State Monopoly. The Soviet Union has usually been free to establish agencies in foreign countries, often employing well-known, experienced domestics of those countries who effectively secure freight for Soviet shipping lines. No such opportunity exists for Western shipping lines to establish effective agencies in the Soviet Union. This has worked to the great advantage of the Soviets, not only in shipping but also in such arrangements as airlines. It has been particularly pernicious in bilateral trade with the Soviets who generally insist on buying FOB and selling CIF, which has assured them the lion's share of the traffic. In the case of British-Soviet trade, it has been reported that the Soviets have dominated the transportation of goods at a ratio of 9 to 1. An FRG government representative reminded Igor Averin, Morflot's Head of Foreign Relations, in Hamburg on October 10 1983 that the Soviet flag was still carrying 65 per cent of the 600,000 tons per annum of liner cargo in the FRG-Soviet trade. Overall, approximately half Soviet foreign trade by sea has been carried by the Soviet fleet, while the British fleet manages to carry only about one-third of the British trade and the US fleet only about 6 per cent of total US trade, although 30 per cent or more of the dry cargo liner traffic.

Unfair Soviet Accounting. Capitalistic shipping

owners and managers correctly contend that Soviet
shipping lines do not have to bear the total costs
of their operations, and therefore they can quote
arbitrarily low freight rates. Many of their costs
are absorbed by the overall Soviet economic system.
These would include such things as capital
investments, amortisation of equipment, and fringe
benefits of crew members, such as health care, and
general subsidisation of such things as housing,
utilities, commuter transportation, and many other
everyday 'social wages' which allow cash wages to
be much lower than those in countries where the
individual has to bear such costs. It has been
stated that the average seaman's wage is only about
one-third that of many Western countries, although
it has also been conceded that Western shipowners,
through registry under flags of convenience,
commonly employ crews from less developed countries
where wages are probably no higher. Much of the
training of Soviet merchant crews is provided free
by naval institutions, and mariners are
interchangeable with naval crews. The Soviet state
assumes all insurance responsibilities, thereby
negating the need for shipping lines to insure their
cargoes, equipment, or personnel (this can also be
said of some large Greek shipping magnates who
remain self-insured). Most fuelling of Soviet ships
is done in Soviet ports where bunkering costs are
usually only about one quarter those in foreign
ports. As has been pointed out, certain other
countries, notably Peru and Saudi Arabia, do the
same thing, but they are not viewed as seriously
because they are not military threats to the rest of
the world. Mitsui estimates that total ship-day
operating costs that must be accounted for by Soviet
shipping lines are no more than one quarter of those
of Western shipping lines.

Reneging on Gentlemen's Agreements. Occasionally
the Soviets have, under duress, informally agreed to
comply with the wishes of trading partners whose
cargo they are carrying as cross-traders, and then
subsequently have blatantly ignored these
agreements. Recently the most extreme example of
this has been FESCO (Far Eastern Shipping Company)
based in Vladivostok. In 1980 it found itself with
excess cargo capacity which could no longer be
unloaded at US Pacific ports. It then asked the
freight conference governing shipping between Japan
and other Far Eastern countries with Australia for

permission to operate as a tolerated outsider. It promised that its liners would be loaded at least 50 per cent with Soviet exports to Australia and that it would fill only residual space with cargo cross-traded from Japan, Hong Kong, and so forth. Two years later the conference came to the realisation that hardly any Soviet exports moved to Australia from Pacific ports and that of the 18,200 TEUs being transported annually probably only about 20 were actually carrying Soviet cargoes. The governments of Australia and Japan have ineffectively called for discussions with the Soviets about this matter, but the Soviets have put them off time and again. It appears that the only way to counteract such devious behaviour is for the Australian unions to take some decisive action such as the US Atlantic Coast longshoremen took in 1980[8].

Current Primary Issues Regarding Soviet Shipping

During the 1970s Soviet shipping lines pursued high-value cargoes on the most lucrative routes, particularly in the North Atlantic in the trade between the United States and Europe and in the North Pacific between the United States and Japan, other Asian countries, and Australia. By 1978 they were operating eleven liner services in the Trans-Pacific and Trans-Atlantic trades of the United States, and more than 90 per cent of the cargo on their ships consisted of goods in United States trade with non-communist countries, thus generating hard currency for the Soviet Union. The Soviets began servicing shipments between the East Coast of the United States and Europe after 1971, and by 1977 they were carrying 12 per cent of the trade between the North Atlantic Coast of the United States and West Germany, 4.1 per cent of the trade between the North Atlantic Coast of the United States and Netherlands-Belgium, and 3.9 per cent of the trade between the North Atlantic Coast of the United States and the Mediterranean-Black Sea.

In addition, the Soviets were carrying 4.4 per cent of the trade between the Gulf Coast of the United States and Britain and the European Continent. In the Pacific in 1977 they were carrying 6.1 per cent of the trade between the Pacific Coast of the United States and the Far East, 12.8 per cent of the trade between the Gulf Coast of the United States and the Far East, and 10.3 per cent of the trade between the Atlantic, Gulf, and Pacific Coasts of the Unites States and Indonesia, Malaya, and Singapore[9]. Moreover by 1974 Soviet ships were carrying 11.2 per cent of the freight between the North American Great Lakes and Europe.

By the end of the 1970s, the USSR had captured about 25 per cent of the cargo on the Europe-East African route, frequently cutting rates 50 per cent under conference levels, and offering 36 sailings per year against the twelve that the conference had considered adequate. A similar situation had developed in the Europe-Central American West Coast trade where Soviet ships apparently carried about 25 per cent of all cargoes. And in 1972 the Soviets began promoting the Trans-Siberian railway as a land bridge alternative to the all-sea route between Asia and Europe. Through offering rates 20-40 per cent below conference rates, the Trans-Siberian was apparently able to capture almost 25 per cent of the Asia-Europe cargo by the end of the 1970s[10].

Much of this significantly changed after the Soviet invasion of Afghanistan in 1979 which prompted US Government embargoes and a longshoremen's boycott of Soviet shipping along Atlantic and Gulf ports of the United States. The Soviets eventually ceased operations in the Pacific ports of the United States also, so that now no American trade is carried in Soviet ships except for that which filters through foreign ports, notably Canadian, primarily Montreal in the east, and to some extent Vancouver in the west.

The cessation of Soviet activity in US trade has released considerable amounts of Soviet shipping capacity which have been deployed in other sea lanes, notably Europe-Far East, Europe-East Africa, Europe-Central America, and Far East-Australia-New Zealand. These shifts have greatly exacerbated existing cut-throat competition on these sea lanes and have further hastened the laying up of excessive capacity, particularly among British ships. As is shown in Table 4.2, there was a decline of 23 per cent in 1976-81 and of another 24.1 per cent during 1981-83 in the UK's active deadweight tonnage. No wonder British shipping interests have been most vociferous in decrying the continuing penetration by the Soviet fleet.

The Atlantic Route. Arctic Line ships headquartered in Murmansk have called at Montreal since 1964. In late 1982 they started a fortnightly outsider container service, Western

Europe/Montreal-Toronto and return. Ice-class ships visit Montreal year round. This supplements an already existing Balt-Atlantic ro/ro service to Montreal. The Soviets expect that there will be a rapid increase in trucking of US cargo both ways across the Canadian border to take advantage of their cheaper Trans-Atlantic rates. These two lines together offer three sailings a month to Hamburg from Montreal with a transit time of approximately eleven days. Trans-shipment services are available to the United Kingdom and Scandinavia. A rate war on the North Atlantic has induced the Soviets to withdraw from calls at Felixstowe and Antwerp on their east-bound trip. It is anticipated that Balt-Atlantic in 1984 will provide a weekly service from north-west Europe to Montreal with an annual TEU capacity in both directions of 99,840. In addition, POL (Polish Ocean Lines) is running new ro/ro ships across the Atlantic, undercutting conference rates by 40 per cent or more.

By late 1984 Balt-Atlantic, Arctic Line, and POL together will be offering 270,000 TEU capacity between north-west Europe and East Coast North America, which will be 10.3 per cent of the total 2.6 million TEU capacity. Balt-Atlantic and Arctic Line with 5.9 per cent of the capacity will be using the St. Lawrence Gateway method of covering US traffic. In addition, Morflot has finalised arrangements to market a Trans-Atlantic container service that will connect with the Trans-Caucasian rail system to Iran. A new conventional service, Trans-Atlantic-north Iran, will commence service from the Great Lakes-St. Lawrence-East Coast ports to Jolfa in Iran via the Finnish port of Kotka during summer and via Bremen and Hamburg during winter.

The Far Eastern Freight Conference. The FEFC was founded in 1879 by a number of shipping lines that were anxious to limit cut-throat competition in cargo carrying between Europe and the Orient, one of the longest and most heavily trafficked sea lanes in the world. The conference now consists of 32 member lines which periodically set agreed-upon freight rates between ports in the United Kingdom, Finland, Sweden, Norway, Denmark, Germany, Holland, Belgium, France, Italy, and Yugoslavia in Europe and ports in the states of Malaysia, Singapore, Thailand, Hong Kong, Taiwan, Japan, and South Korea in the Far East. In opposition to the FEFC, Morflot now offers its longest cross-trade enterprise, between north-west Europe and the Far East, with a weekly service by ten modern ships. This involves two Soviet shipping companies, the Balt-Orient Line from Leningrad on the Baltic and the Odessa Ocean Line from Odessa on the Black Sea. In competition with both FEFC and the Soviet shipping lines is the Trans-Siberian Container Service, which has been mentioned earlier.

Recently, both the shipping routes and the Trans-Siberian route have been thrown into chaos by two independent shipping lines from Taiwan, Evergreen and Yang Ming. They have induced a rate war which has resulted in plummeting conference rates that now for many commodities are somewhat below the Trans-Siberian rate. The shipping lines in the conference have vowed that they will fight Yang Ming to the end, so further rate reductions might be expected. And in April, 1983, the COSCO shipping line from the People's Republic of China proposed to start a fortnightly container service to Europe. However, nothing has been said about this since.

According to Lloyd's List for November 3 1983, the Soviet carriers, all services taken together, are clearly the biggest culprits in the rate war on the Europe-Far East route. The Balt-Orient line, part of the Baltic Shipping Company of Leningrad, and the Odessa Ocean Line of Odessa between them have offered 67 per cent more container space to the Europe-Far East trade during the first 10 months of 1983 than the same months of 1982. While Yang Ming increased its share of Europe-Far East annual capacity both ways from 1.7 per cent in 1981 to 3.4 per cent in 1983, the Soviet all-water lines have expanded their slice of the total market from 3.0 per cent in 1981 to 6.5 per cent in 1983. The Soviet all-water slot deployment will have risen from less than 49,000 TEUs in 1981 to more than 123,000 slots by 1984/85. Consequently, the overall container space, including the Trans-Siberian, offered by the USSR in the Europe-Far East trade, excluding Orient-Iran capacity westbound, will rise from 76,000 TEUs in 1977 to more than 191,000 slots by 1985. This compares with an increase by Yang Ming from 26,440 TEUs in 1981 to an estimated 79,378 container spaces in 1984.

Anti-communist inspired loading restrictions in Taiwan and South Korea have restricted the Soviet all-water activity to Jeddah, Singapore, Thailand, Malaysia, the Philippines, and Hong Kong, while the Soviet lines cater for the Japanese market via the

Soviet Shipping : Its Impact on the West

Trans-Siberian Land Bridge.
    During 1983 the Soviet lines deployed a growing
number of their new, bigger 960-TEU class to provide
sailings at approximately ten-day intervals.
Balt-Orient Line deployed seven 825-TEU vessels and
four 960-TEU ships on slightly better than weekly
average frequency. It appears that the Odessa Ocean
Line may be able in 1984 to mount a fortnightly
express fully cellular service based entirely on the
new class of 960-TEU vessels between the Black Sea
and the Far East. Although Odessa claims that its
purpose is to service Black Sea ports with South
East Asia, all of its fully cellular ships make a
lengthy detour west to Genoa after leaving the Black
Sea before proceeding eastward through the Suez
Canal to the Orient, thus indicating that the Odessa
Line is carrying considerable Italian export cargoes
to the Far East.

The   Trans-Siberian   Container   Service.   The
Trans-Siberian   Container   Service   (or   the
Trans-Siberian Land Bridge) came into operation in
1972 as an alternative to the all-water route from
the Far East to Europe, which at that time was going
all the way around the southern tip of Africa
because the Suez Canal was still clogged by sunken
ships. Even with the Suez route open, the distance
from Yokohama to London is more than 22,000
kilometres, as compared with only 13,000 kilometres
via the Trans-Siberian. The primary originators of
traffic westbound are Japan, South Korea, and Hong
Kong. The Soviets hope to extend services to the
Philippines and Australia. Until 1983 Nakhodka was
the primary transit port for containers in the
Soviet Far East, but now it has been announced that
Vostochnyy is the sole transit port for the land
bridge, and Nakhodka has been reserved for domestic
and foreign trade.
    By undercutting conference rates, and offering
comparable transit times, the land bridge steadily
increased its container cargo until 1981 when it
carried a record 103,843 TEUs both ways. 45,060 of
these were westbound to Europe, 37,747 westbound to
Iran, and 21,036 eastbound. In addition, there
probably were more than 20,000 eastbound empty
containers, since there is such an imbalance of
container movement in the two directions.
    However, after its initial success, the
Trans-Siberian suffered a large drop in cargo in
1982 due to import curbs imposed by Iran on its

indebted economy, brought on by the Iraq war and
continuing chaos in the aftermath of the revolution,
the economic recession in Western Europe, and
reduced shipping rates on the all-water routes.
Between April 1982, and March 1983, the
Trans-Siberian carried a total of 63,821 containers
westbound, about 42,000 to Europe and about 21,800
to Iran. Containers eastbound totalled 22,842, of
which 15,750 were empty[11]. Exports from Japan,
Korea, and Hong Kong using the land bridge decreased
by 11.5 per cent in 1982. Cargo volumes from Japan
to Europe fell 6 per cent. Total carryings on the
Trans-Siberian fell from 1.1 million tons in 1981 to
980,000 tons in 1982. Speedbridge, a leading
Japanese consolidator for the Trans-Siberian route
who recently went bankrupt, reported that during
1981 about 40 per cent of Japanese home electrical
appliance exports to Europe went via the
Trans-Siberian, but in 1982 this slumped to only 4
per cent. It appears that during 1983 westbound
traffic to Europe remained static while westbound to
Iran set a new record of about 45,000 TEUs.
Eastbound traffic appears to be up a little, but
that is not particularly significant, since
eastbound movement is primarily to reposition empty
containers.

In addition to freight rate reductions by
shipping lines, the Trans-Siberian route seems to
have run into some other problems. Delays along the
route have lengthened the average transport time for
a container from Japan to Europe to an average of
about 30-35 days, as opposed to 18-25 days by the
all-sea route, and in some cases it is reported that
a container can take as long as 90 days via the
Trans-Siberian. Such delays have resulted primarily
from the lack of containers at the eastern end as
well as a shortage of service trucks which,
particularly during harvest, are routinely
commandeered to haul grain. In addition, it has been
reported that much pilfering is taking place at rail
sidings. Also, because of strained political
relations, American-Iranian trade via the Pacific
and Trans-Siberian route has ceased.

The Soviets are trying to improve their
service, and hope that after the rate wars end they
can greatly increase transit freight on the land
bridge. In preparation for this they are rapidly
completing many construction projects related to the
Trans-Siberian route. Not the least of these is the
BAM (Baykal-Amur Mainline railway) which was originally scheduled for completion by the end of 1984.

Soviet Shipping : Its Impact on the West

The BAM will provide an alternative and shorter route along the eastern half of the land bridge and relieve some of the overload of traffic on the Trans-Siberian which probably slows up train movements. But it will do nothing for the West Siberian portion of the land bridge, which is the most heavily trafficked part of the Trans-Siberian railway, nor for movements through the European part of the country. The Soviets are hoping to solve some of this problem by electrifying the entire route and using longer block trains pulled by more powerful engines to increase speeds. A pilot container was moved from Vostochnyy to Leningrad in eleven days. The Soviets hope that in the near future they can make this the norm rather than the marvel[12]. The Soviets report that electrification is nearly completed all the way from Brest on the western border to Vladivostok in the Far East.

No less important are the improvements being made in the handling of containers at either end of the land bridge. In 1984 apparently there was still only one container terminal operating in Vostochnyy, with a handling capacity of 150,000 TEUs per year. But within a year or two the Soviets hope to have two more terminals of similar size in operation, and some time in the future to add five more, which will bring the total capacity to 600,000 TEUs per annum. Thus, there will be no lack of handling capacity at the eastern end. At the western end many variants, rail, road, and water, have always been used to carry freight west of the Volga River to Eastern and Western Europe, the Mediterranean, and Iran and Afghanistan. Primary improvements under way at present include the following: (1) development of an easy rail-truck transfer at Brest for the ongoing transport of containerised freight directly to many European destinations; and (2) a system of train ferries now being set up in the Baltic to convey containerised freight from Klaipeda to Kiel in West Germany and from Klaipeda to Mukran-Sassnitz in East Germany, thus bypassing troubled Poland. Both of these routes are to begin operation in 1986. (3) Also, the Soviets report that they are in the process of building the largest container port in the Soviet Union at Tallinn.

In addition to these various construction projects, the Soviets are introducing computerised container tracing, which has been a major problem. And they have hired the West Germans to build a modern container manufacturing plant at Il'ichevsk

138

near Odessa,which is now turning out about 5000 TEUs per year. Eventually they hope to phase out their time-chartering of containers from foreign companies.

To make up for some of the loss of container movement between the Far East and Europe, the Soviets have extended services to China and plan in the near future to extend them also to Mongolia. Jeuro Container Transport, based in London, who has been handling about 35 per cent of the westbound container traffic on the Trans-Siberian route and 52 per cent of the eastbound, established the Sino-Siberian route and set up its own office in Peking. During 1982 it handled 2000 TEUs from Mainland China and during 1983 about 5000. It hopes that within four years the China trade along the Trans-Siberian will equal the present Far East-European trade. Jeuro has moved boxes from China to the United Kingdom in 28 days, which is about the same transit time as the deep sea route. The Chinese are not set up to handle containers efficiently, and Jeuro is picking up cargo at inland points. Anything in the northern half of China goes cheaper by rail through Siberia to Europe than around the southern water route. Jeuro operates out of three UK ports to the Far East via the Trans-Siberian Land Bridge. Vessels sail from Tilbury and Hull to Leningrad and from Ellesmere Port and Dublin to Riga. Sailings are weekly from Tilbury and fortnightly from the other ports.

Far East-Australia. The case of FESCO (Far Eastern Shipping Company) as a tolerated outsider by Japan and Australia has already been cited. For a number of years the Soviets stayed within a generally recognised fair rate reduction below conference of about 10-15 per cent, but during the past year they have sometimes cut rates as much as 50 per cent. Also, of course, they have reneged on their agreement that half the cargo be Soviet exports to Australia and New Zealand. The previous Australian government stated that it had no intention of using its influence to deny the cheapest rates possible to Australian shippers, but it did emphasise that it would be concerned if non-commercial competition from state-owned shipping lines threatened to destabilise and risk the guaranteed future of services to ports around Australia. The new Australian Labour Government is giving greater attention to demands from maritime unions for job security within the Australian shipping industry, and they are more committed to the continued operation of the government-owned Australian National Line, which has been one of the hardest hit by FESCO. The Australian Council of Trade Unions has warned FESCO and also ZIM and Hong Kong Island Line 'to change the present situation or face regulation by the ACTU itself.'

Cruise Ships. The Soviets have entered heavily into the most lucrative shipping trade, the cruising business. In 1982 Soviet liners carried 3.2 million passengers, including 400,000 foreign passengers on cruises outside the Soviet Union. They have been most active in the European-Mediterranean trade, but also have cruise ships running across the Atlantic from Leningrad to Montreal. Altogether Morpasflot operates more than 70 cruise liners.

During the period 1981-3 their share of berths offered in the UK market by all cruise companies has risen from 14 per cent to 26. Their fares have been as much as 40 per cent cheaper than others. This has elicited much complaint from cruise companies in the United Kingdom, which finally resulted in a pledge from Morpasflot to reduce its cruising berths in 1984.

A similar situation evolved in Italy where domestic tour operators chartered Morpasflot ships on 68 cruises during 1983 that amounted to more than 40 per cent of the Italian market. Italian tour operators have agreed to cut cruising days using Soviet cruisers from 250 in 1983 to 200 in 1984.

Australia has put an absolute ban on Soviet cruise ships in its area, which, incidentally, has put in jeopardy a large contract to sell canned fruit to the Soviets.

EEC-Soviet Negotiations. The European community has become so annoyed at Soviet refusal to limit their operations that European governments are considering protectionist quotas and/or equalising taxes against Soviet shipping. The European countries are particularly concerned about the revelation that some EEC aid cargoes to the Third World are being carried by Soviet ships. The Soviets have responded that their trade organisations are big freighting customers for cargoes under foreign flags, and if provoked this could change. Finally, in 1982 the Soviets for the first time agreed to

140

some concessions. They agreed to a 30 per cent reduction in the volume of coffee carried between Central America and Western Europe by Soviet ships, and on the same run that cotton shipments would be limited to one-fifth of the total market. Also they agreed to a 10 per cent reduction in the amount of pickups in Western Europe by Soviet vessels for shipment to Africa. However, the Soviets did not agree to any price increases. The EEC has agreed to a monitoring system of cargo rates carried by Soviet ships between Europe and the Far East and Europe and Africa. However, it appears that only the United Kingdom has initiated monitoring.

Summary

It appears that shipping companies and ship builders world-wide are in for a long-continuing depression. Shipping capacities are far overbuilt, and trade has stagnated. State-owned lines, such as the Soviet ones, have the advantages of large state subsidies and lack of cost accounting, so they can set rates arbitrarily low. Costs are also lower in many of the Third World countries which are developing rapidly expanding fleets. Competition from such outsiders has induced conferences to undergo successive rate reductions, which is causing lines to operate at a loss. In some instances, even the Soviets have been reducing their operations. So far the Soviet merchant fleet has handled less than seven per cent of world trade, and therefore cannot be considered a primary threat. But within the context of the extreme crisis that exists everywhere in world shipping, even this small interjection by an unyielding competitor has caused much consternation among established shipping lines.

NOTES

* The author gratefully acknowledges the co-operation of Ken Moore, General Council of British Shipping, who supplied a constant stream of information, and Mary Lydolph who typed the manuscript.

[1] A Review of Developments in World Trade and Their Effect on the Shipping Market, World Trade Review and Outlook No. 26 (Lambert Brothers Shipping Ltd., London, March 1983), 42 pp.
[2] David Scrivener, 'Merchant Marine in Soviet Naval Strategy', Marine Policy, 7, 2 (April

1983), pp.118-122.

[3]   Ibid., p.119.
[4]   G. Smalley, 'Role of Merchant Fleet in Soviet Global Strategy', Marine Policy, 8, 4 (January 1984), pp.65-8
[5]   'Herbert Fromme reports from the Third International Liner Symposium in Bremen', Lloyd's List, (November 3, 1983).
[6]   Smalley, Role of Merchant Fleet, p.66.
[7]   Scrivener, Merchant Marine, p.120.
[8]   For a fuller account, see 'A case of Soviet Skullduggery', in The Challenge of Soviet Shipping, (Aims of Industry, London, 1983), p.29.
[9]   John P. Hardt, 'Maritime Developments Involving the Soviet Union, the United States, and the West', in Issues in East-West Commercial Relations, (Joint Economic Committee, Congress of the United States, Washington, January 12, 1979), p.252.
[10] Hendrik van Rijn, 'The Trans-Siberian Railway: Already a Looming Threat', in The Challenge of Soviet Shipping (Aims of Industry, London, 1983), pp.30-6. Other sources give considerably lower estimates.
[11] Shipping and Trade News (June 21, 1983).
[12] van Rijn, The Trans-Siberian, p.32.

GLOSSARY

Bulk carrier : a large dry cargo tramp ship, conventionally of over 18,000 deadweight tons
C.i.f. : ('Cost, insurance, freight') a price which includes all expenses up to the port of import
Conference : a grouping of shipping companies engaged in trading between two areas, who agree to charge standard rates for individual commodities, and who usually limit the number of ships engaged in such trade so as to give shippers a reasonable but not over frequent number of calls in a period; the objective is to reduce the fluctuations in price resulting from an inelastic supply of ships
Cross trade : freight carried by a ship registered under a flag of a country not called at during the voyage
Deadweight ton : the maximum weight in tons of cargo, fuel, water and stores, which a ship is able to carry, measured by converting the difference between cubic feet displaced when fully loaded, and when empty

Gross ton : a measure of a ship's size, taken as equivalent to 100 cubic feet of permanently enclosed space

F.o.b. : ('Free on board') a price which includes all costs up to and including handling charges to load the freight on a ship

LASH : ('lighter aboard ship') a ship which carries lighters, or barges; the 'mother' ship can load and discharge these barges in deep sheltered water, without berthing in a port

Liner : a ship with a periodic and regular schedule between given ports, carrying the freight of many shippers on any one voyage

Ro-ro : ('roll on, roll off') a ship designed to carry wheeled vehicles (including rail freight cars) and trailers, in which the cargo is itself carried

TEU : 20 foot (container) equivalent unit, used to express traffic in comparable units; a 30 foot container = 1½ TEUs, a 40 footer = 2 TEUs

Tramp : a ship, not used for a regular service, carrying in the main homogenous dry cargo in bulk according to demand; such ships are chartered on a time or voyage basis

Chapter 5

SOVIET AIR TRANSPORT: GEOGRAPHIC, TECHNICAL AND
ORGANISATIONAL PROBLEMS

Leslie Symons

Any attempt to assess quantitatively the efficiency
of Soviet operations in the sphere of civil aviation
is severely limited by the absence of published
statistics and the virtual impossibility of
extracting any unpublished information from any
source in the Soviet Union. Whereas in Great Britain
one could pursue enquiries relatively unhindered and
could expect access to many of the records of the
Civil Aviation Authority, airports authorities and
many of the airlines, subject only to reticence on
some commercially sensitive aspects, no equivalent
material is available in the USSR. The difficulty of
obtaining statistical information beyond the modest
amounts published in the official statistical
handbook is, of course, familiar to all who work on
the Soviet economic scene. In the case of aviation,
however, the problem is further bedevilled by the
apparent assumption in Soviet government circles
that information relating to all aspects of the
aviation industry has some military significance and
that therefore only a bare minimum, to demonstrate
the undeniably impressive achievements of the
airlines, should be revealed.

Compared with the statistics available for
agriculture, for example, those published for Soviet
aviation may be likened to the struggling plants of
Novaya Zemlya compared with the lush vegetation of
the Crimea. Even in the slimmer volumes of Narodnoye
khozyaystvo SSSR to which we have become accustomed
in recent years, the agriculture section runs to
about 90 pages (excluding relevant material
elsewhere in the volume) whereas all transport
matters get only about one third that number of
pages while aviation is accorded just two pages.
The volumes appearing for the individual republics
are similarly barren. The occasional volume of

144

Transport i svyaz' is only marginally more
informative on air transport. Virtually nothing is
published on the aircraft construction industry,
except data on individual types of aircraft such
as appear in the annual volumes of Jane's All the
World's Aircraft, an indispensable source, and the
data sections of magazines catering for the aviation
industry, notably Flight International. These
sources we will return to later; let us first
consider the officially published statistics.

Narodnoye khozyaystvo tables present figures
for the annual totals of passengers carried,
passenger-kilometres (passazhiroborot = passenger
turnover or total traffic), metric tons of freight
carried, freight turnover or total traffic, the
overall length of air routes, and overall
coefficients of passenger and freight loading of
aircraft. Agricultural and forestry operations,
mainly application of fertilisers, pesticides and
defoliants, are given by republics. This, then, is
all the raw material available annually. Editions
of Transport i svyaz' add details of agricultural
work by type of operation and some regional
information on passenger numbers but this is very
little and the series is added to only at rare
intervals.

Apart from these sources there are two regular
publications for the Soviet aviation world, the
monthly magazine Grazhdanskaya aviatsiya and the
newspaper Vozdushnyy transport, which is published
three times each week. The former has the more
general articles, including some coverage of new
types of aircraft, etc., with colour photographs
and diagrams and probably has a considerable
readership among the members of the Soviet public
interested in aviation. It also has fairly
technical articles on operating procedures,
ancillary equipment such as radio, etc., together
with the politically inspired reports and
exhortations common to all Soviet papers. Vozdushnyy
transport is a broadsheet of four pages per issue,
therefore looking like an ordinary newspaper, and
commonly having leader articles and pictures
devoted to party activities, congresses and the
like, with aspects of special relevance to civil
aviation, such as national operating targets for
Aeroflot. Most of the space is, however, taken up
with matters of interest to career aviators, such as
reports on the introduction of new equipment at
airports, training requirements and seasonal
problems, such as overcoming winter hazards and

summer overcrowding at airports, together with trade union and legal topics.

Unfortunately, as in other Soviet publications, there is a lack of hard statistical material usable to make effective assessments of performance, and no details are given of networks, fleets or accidents. However, as in other Soviet papers, space is given to criticism from individuals holding particular offices such as airport manager and sectional heads as well as line pilots. This material gives interesting insights into the problems of maintaining efficient operations, the pressure for economy, disputes between the representatives of different ministries, etc. From these it is possible to build up some sort of overall picture of the kind of problems that beset the civil aviation industry in the USSR and the points of strain which appear from time to time or persistently.

To return to serials published in the West, these bring together material supplied by the industry - aircraft firms, airlines, etc., and, particularly in the case of Flight International and other magazines, reports from a variety of sources including Vozdushnyy transport and Grazhdanskaya aviatsiya. Such abstracts are, however, few and brief. The main value of Flight comes in its collection of news items from any available source, while its special issues, like Jane's, provide collected data in detail. Articles and news items of relevance appear from time to time in Airports International, Interavia Aerospace Review, Aviation Week and Space Technology and other journals.

The principal indicators of air transport performance in the USSR are given in Table 5.1. These show the rapid growth in the importance of this mode of transport and the slowing down in growth in recent years. After such rapid growth some slowing down is to be expected and world economic problems have exacerbated this predictable trend. This decrease in growth is, however, a cause for concern in the USSR and has occurred despite the fact that the Soviet Union offers many advantages for air transport. Firstly, its great size makes aviation a particularly attractive method of movement, saving at least a week on the longer routes in comparison with rail transport. Secondly, its climatic regimes, although very varied and harsh in comparison with those of maritime countries in temperate or even tropical latitudes, are largely favourable to air movements, or at least offer fewer obstacles than they do to surface transport.

Thirdly, the size of the national economy is sufficient to support a full range of aeronautical industries. This is particularly important because of the high level of investment needed for aircraft and aero engines. In addition, Soviet command of the Comecon bloc provides it with a captive market for exports, and guaranteed extensions of routes for its airlines. Comecon, which includes Mongolia, Cuba and Vietnam as well as the six Eastern European countries has a total population of 450 million (275 million in the USSR) and accounts for about one-third of the world's industrial output.

Where then lie the problems? Paradoxically, all the above advantages contain within them disadvantages which lead to organisational and cost problems.

It may also be argued that the socialistic form of organisation, whereby the state has a monopoly in the provision of public service transport, is also an advantage, at least in providing services. It avoids cut-throat competition and can· remove uncertainty concerning the survival of the less heavily used routes where costs per seat-kilometre are high. The needs of the state are paramount and aviation is vital to the survival of the state, so it is unlikely to be allowed to suffer from want of finance. Nevertheless, it is clear that the USSR experiences problems in the design and production of new types of aircraft, in the organisation and administration of Aeroflot, and in supply matters which may affect not only the reliability, but also the safety of its operations. Each of these aspects of Soviet air transport are considered below.

The Size of the USSR - Costs for Air Transport
At over 22,400 million square kilometres, the Soviet state is by far the largest in the world. Inevitably, it includes vast areas of thinly populated terrain, negative because of climate and relief, which can support only the most skeletal of ground transport services, so that the burden of supplying them falls to a very large extent on the airlines. This requires the provision of a vast number of airfields which are relatively thinly used and therefore expensive to maintain, together with the ancillary services such as navigational aids, fuel transport and engineering support necessary to operate over these territories.

Even on main routes between important cities, the almost uninhabited areas east of the Ural mountains require chains of navigational aids and maintenance of adequate diversionary airfields for use in emergency. These costs are, of course, borne by the state and can be shared with the armed forces, but the location of their support services and, particularly, airfields with the required runway lengths, will not necessarily coincide with the needs of the civil airlines.

In addition there is the problem of production of the vast quantities of aviation fuel required. The transport of fuel to distant airfields where local operations must be based, and where long distance aircraft may regularly be required to refuel, is particularly expensive. The larger aircraft can carry fuel for very considerable ranges, including nonstop flights over the longest routes even the USSR requires for its internal and international services, but refuelling should normally be carried out before return flights. Even where this is not essential, because of the range of the aircraft provided by its own fuel tanks, it will be economically desirable because carrying fuel for longer distances than necessary reduces payload. Hence it is desirable to transport fuel to refuelling points by pipeline, waterborne tankers or rail, but airports are often substantial distances even from railheads, while only a few are near ports or pipelines.

The amount of fuel for which provision is necessary is indicated by the capacity of current service types. The Il'yushin Il-86, the largest aircraft in service with Aeroflot in 1984, has capacity for up to 64,000 kg (80,000 litres, 17,600 Imperial gallons). Even the medium-range Tupolev Tu-154 carries 39,150 kg (approximately 49,000 litres) for normal flight, giving a range of up to 4,000 km with 120 passengers. On shorter routes the normal payload is 152 passengers in summer, 144 in winter, illustrating the payload gain when less fuel is needed for the flight. An additional tank carrying 6,600 kg (8,250 litres) of fuel is fitted but is not available for use in flight. This tank can be filled when the aircraft is required to carry less than a full payload and its contents transferred to the main tanks at the destination or a staging airport. This can be used to reduce purchases of fuel outside an operator's home country but, more importantly perhaps in the USSR, it can limit the demand for fuel at airports where supply is not easy.

Climate and Relief.   The USSR has many variations
of climate within its boundaries. It may be said
that, in general, the climates of the Soviet Union
are relatively favourable to air transport because
of the low percentages of cloud cover found in this
great continental area compared with, for example,
Western Europe. Summer conditions are usually good
with only occasional incursions of moisture-laden
air masses when depressions move in from the Baltic
and Black Sea areas. Convectional storms are
frequent but tend to be relatively localised and
of short duration, though lightning always poses a
hazard to aircraft. Dust storms present hazards in
southern areas, especially to aircraft landing. In
general, however, in all except the Arctic regions,
the climates are favourable in summer.

Winter conditions are more mixed. The Siberian
high-pressure system ensures that the incursion of
moist airstreams is not common so that much of the
flying is in clear skies with light winds. On the
other hand, temperatures fall to very low levels,
with large areas below freezing for several months
for much, if not all, of each day. Under these
conditions, maintenance of aircraft and engines is
difficult and it is very expensive to provide
adequate hangarage for servicing of large aircraft.
Starting of piston engines is rendered difficult,
especially where damp is combined with low
temperatures. Turbine engines, however, power most
modern aircraft and these are less difficult to
start. De-icing systems on aircraft must be of the
highest order but these are in any case necessary
for operation at the high altitudes now normal.

Clearance of snow and ice from runways is a
constant necessity throughout the central and
northern regions of the USSR, and even in Central
Asia equipment is needed for the lesser but frequent
falls of snow experienced in the winter months. The
USSR has lacked sophisticated equipment to deal with
this problem until recently. Reliance has tended to
be placed on robust but simple machinery and a large
amount of labour. Snowblowers of modern design have
been supplied to an increasing number of airports in
recent years, but it is unlikely that enough of
these machines are available at present for them to
be a significant factor. In the UK a large
snowblower costs over £100,000 so supply of this
type of machine to the large number of airports in
the USSR requires considerable investment.

Ice frequently occurs on runways irrespective

of snow cover. Detection of freezing of the surface moisture is vital for safety and busy airports in the Western world have installed special sensors for this purpose. This practice is spreading rapidly. For example, London Gatwick has (mid-1984) 4 sensors in operation. Heathrow has had none in the past but is now being equipped with them in accordance with a decision to install sensors at all airports controlled by British Airports Authority. Some municipal airports, such as Manchester, have similar installations. Whether such sensors have been developed in the USSR is not known.

Although compared with many parts of the world the Soviet climates are relatively stable, very severe storms are not uncommon. Of particular threat to all forms of transport, but especially to aircraft landing, are the blizzards that sweep across northern and central regions. As a result aircraft are commonly routed to diversionary airports (though even this can be difficult in widespread storms), and once on the ground may be immobilised for days. Major airports are not infrequently closed for several days. Severe storms occasionally close airports even in summer, as in the case of Moscow airports in August 1980. On this occasion thousands of families and school children returning from holidays were held up and special measures were taken to provide for their food and rest (Voz. Trans. 4 Sept. 1980). The very next month saw Moscow airports, especially Vnukovo, affected by closure of airports in northern areas for several days and again special provision had to be made for delayed passengers (Voz. Trans. 9 Sept. 1980). This is an example of the way bad weather in one area affects other regions. Late autumn brings increasing likelihood of severe disruption. In November 1981 bad weather forced closure of Ukrainean airports on 315 occasions, totalling 2,216 hours spread over all but four of the days of the month (Voz. Trans. 3 Dec. 1981).

The approach of winter is regarded as a period for intensive overhaul of all forms of equipment and of special training sessions for staff, especially aircrew. The need for maximum effort to prepare for all emergencies is stressed in special notices to staff. It is also, however, noted from time to time that there is inadequate equipment, or lack of training of crews to handle the equipment to full advantage. An example of the recognition of special risks to some types of operation is the attention given to aircraft on short-distance (mestnyy) routes. Such aircraft fly at relatively low levels and therefore are not able to stay above the weather for long periods in the way that is possible for long-haul flights. Also, the smaller aircraft usual for local operations do not carry such specialised and sophisticated aids and more reliance is placed on visual flight operations so such services are more frequently disrupted by bad weather (Voz. Trans. 25 Sept. 1980).

Relief and related climatic regimes are not major problems in the western parts of the USSR, where the population and industry are densest, but in the Caucasus regions, Central Asia and over much of Siberia there are difficulties which can only be overcome at considerable cost. In Central Asia the problems of the 'hot and high' airport are encountered. Owing to thinness of the atmosphere with increasing altitude, aircraft require more thrust and lift to achieve take-off with a given load. To compensate for this it is normal to reduce the amount of fuel and/or payload taken aboard. If fuel requirements in relation to range necessitate full tanks at take-off then the aircraft may have to fly greatly reduced passenger or freight loads. To minimise waste and disruption of services, the Yakovlev Yak-40 trijet was designed for medium and short-haul services with a capacity to take off from short runways. Runway length requirement is an important factor in mountain areas where land flat enough for airports is limited and construction costs of levelling in solid rock are very high. A Yak-40, with full load, requires a take-off run under normal conditions at sea level of 650-700 m, and a field length of about 1,100 m, about half that of older generation jets available, such as the Tupolev Tu-134. Such aircraft were larger, but economical short-field capability is more important than capacity in the mountain regions. This can be seen by the requirement of the Yak-40 at the same temperature at 1,530 m (5,000 ft) for a field length almost 40 per cent greater while an air temperature 25°C higher adds to field length requirement a further 28 per cent. Under these conditions the take-off run of the Yak-40 is still less than that of the Tu-134 at sea level in optimal conditions.

The USSR does not, however, have available aircraft of the more economical kind powered by turbo-propeller engines specially designed in recent years for short-field operation in the West, such as the Short 330 and 360, and De Havilland Canada Dash 7. The Soviet An-28 is intended to fill such a role but as a direct derivative of the ageing An-24 cannot be expected to be as efficient and economical as newer types.

At the other extreme, the swamps and land liable to flood which account for vast areas in West Siberia also provide problems for airport construction. In this case runways may have to be floated on rafts to enable them to be extended over swampy ground. All constructional work in the cold, frosty and dark conditions of the winter half of the year is difficult and costly. The position is further complicated in northern and eastern Siberia by the widespread presence of permafrost below the runway. Seasonal thawing of the upper or surface layers may cause buckling of runways.

The Size of the Economy as a Base for the Aviation Industry

Because of the complexity and cost of development programmes, only the USA and the USSR can support complete national airframe and aero-engine industries capable of producing the most sophisticated of modern aircraft, whether these be large transport aircraft or front-line fighting machines. The USA also has very large export markets, and in order to protect these enters into co-operative ventures with other western countries, particularly in jet engine development, in the expectation of obtaining contracts for powering foreign-built aircraft. Combined domestic and export markets provide the USA and, to a lesser extent, the USSR, with a demand for aircraft and spares far in excess of that generated by even the largest European co-operative programmes, without the further handicap of the complexities of having design teams split between several nationalities and being dependent on several sources of finance, both government and corporate in origin.

The USSR lacks the command of worldwide export markets enjoyed by the USA; its products do not enjoy a sufficiently high reputation to be able to overcome the advantages for customers of being able to deal with the varied sources of supply in western countries, whether American, European, or for some smaller types, other countries as well. Countries which have purchased Soviet aircraft and then wished to change to American or European types have found difficulty in disposing of the Soviet-built fleet, and their abandonment of reliance on the Soviet bloc for new aircraft has further depreciated

Soviet-built aircraft.

The USSR is large enough to require sufficient aircraft of various types to sustain its industry, but the captive market existing in the other Comecon countries is valuable in providing it with a useful addition to the demand for its products, and contributing to its technical requirements. It should be noted that the Eastern European market is not entirely captive. For example, Romania builds BAC 1-11 airliners under licence from British Aerospace, as well as the Islander short-haul light transport aircraft developed by Britten-Norman in the Isle of Wight, but now owned by Ciba-Pilatus of Switzerland.

These and some other exceptions for light aircraft do not, however, invalidate the general statement that the Comecon countries find it necessary to rely mainly on Soviet designs, in the construction of which they are able to participate to a limited degree. Thus the Polish firm, PZL, is entirely responsible for the production of the An-2 biplane for local services and agricultural purposes, and also builds wing sections for the Il'yushin Il-86 airliner. The Czech L-410 short-haul transport has been chosen as a replacement for the An-2 in the current Soviet re-equipment programme.

To what extent such contracts indicate a bottleneck in the Soviet aircraft industry is not clear. Certainly, the placing of contracts for aircraft of a non-strategic kind in Comecon countries helps them to feel that they are being encouraged, or at least permitted, to maintain their own aircraft industries. Being able to rely on these firms for a large number of the less sophisticated machines also means that the Soviet effort can be concentrated on the more important ones. As there is a constant complaint from Soviet government circles that there is a shortage of labour, especially of skilled labour, in the country, it must be presumed that it is also advantageous in this respect to have this contribution from Eastern Europe to the production of certain classes of aircraft and engines.

'Despite the large guaranteed market, precise knowledge of what the users require and obviously almost unlimited backing from state funds, the Soviet aerospace industries have found it difficult to keep pace with the West. It may be that this is largely due to the isolation of their design and production teams from the competitive atmosphere of Western countries. Also, it is likely that not

having the individual demands for particular
requirements from various customers, who are not
only able to go elsewhere but are actively wooed by
other firms, also has a severely stultifying effect.
All this appears to be true not only of the aviation
industry per se but of the essential supporting
equipment and machinery, computers being a
well-known case.

Problems in Soviet Aircraft Design.   Whatever the
causes, the lag behind the west has increased in
recent years. Early in the jet age, as in the space
programmes, the USSR appeared to be highly
competitive and innovative. It is true that it
depended for both programmes in some measure on
foreign technology. The jet aero engine was
developed from the patent stage to operational
success in Europe. The first patent was by Whittle
in Britain, and this eventually resulted in the very
successful Rolls Royce and Bristol engines which
powered such aircraft as the Comet, the world's
first jet airliner, and, eventually, Concorde. During
the war, however, Germany had taken the lead and had
introduced the first jet-powered fighters. British
technology provided the basis for American jet
engines, exported to the USA to facilitate fighter
production, and for Soviet engines following the
supply of Rolls Royce engines under British
programmes designed to facilitate the Soviet war
effort. Once, however, the basic technology had been
acquired, the USSR was not slow to apply it. The
Tupolev Tu-104 was the second jet airliner in the
world to come into service, and the Tupolev Tu-114,
utilising turbine-driven propellers, was for many
years the largest and longest-range civil aircraft
in the world.
When the Boeing 707 and Douglas DC-8 came to
dominate the long-haul air routes of the rest of the
world in the 1960s, Aeroflot was reasonably well
served by designs which were, if not as advanced as
western machines, well adapted to its requirements.
The Il'yushin Il-62 resembled closely the Vickers
VC-10 and remains the backbone of the Aeroflot long
distance routes, long after the VC-10 was ousted
from British Airways by the more economical Boeing
types. Similarly the Tu-154 trijet will undoubtedly
outlive the Trident in first-line service, though it
may not be technically superior. The Soviet designs
have the advantage of not having to provide the keen
economic efficiency or to cope with the stringent

noise regulations which have come increasingly to dominate re-equipment plans in the West.

However, it is in the 'jumbo' and 'airbus' categories, which became so important in the western world in the 1970s, that the Soviet industry has appeared to be markedly behind American and European development. The Boeing 747 design was announced in 1966, first flew in 1969, and made its first commercial flight in January 1970. The first version carried 500 passengers and it has been developed to carry up to 660 passengers, while 'long range versions, freighters and other variants are in service. At least 15 airlines have purchased one model or another and in all 613 Boeing 747s had been sold by October 1983. No comparable aircraft has been brought into service in the USSR, but the need for a very large transport aircraft for probably both civil and military purposes has evidently been felt because in the autumn of 1983 the Antonov An-400 was reported to be flying. Once again the USSR has the world's largest aircraft which will be able to fulfil military requirements similar to those of the US Air Force's heavy-lift Lockheed C-5A Galaxy.

At the next stage down, the Soviet Union also failed for many years to produce aircraft comparable to the Douglas DC-10 and Lockheed L-1011 Tristar. Both of these designs also originated in 1966. The first DC-10 entered commercial service in 1971 with the Tristar entering airline service the following year. The thinking that lay behind these designs for aircraft capable of carrying 380 passengers economically over medium ranges (approximately 500 to 10,000 km) was that airports were becoming heavily congested in terms of aircraft movements and that this could be alleviated by having large machines for moderate as well as very long flights, and capable of operating from airports without especially long runways, e.g. 2,000 m rather than in excess of 3,000 m, as required by the Boeing 747.

Not until 1980 did the USSR bring into service an aeroplane in this class. The Il'yushin Il-86 was chosen after a competition with designs from the Antonov and Tupolev design bureaux. It was first announced in 1971 so that it took at least nine years to design and get to service stage, at least twice as long as the American designs mentioned. The Il-86 carries 350 passengers with a design range of 3,600 km or 4,600 km with maximum fuel, but reports in 1981 suggested that these ranges were not being achieved (Taylor, 1983).

The Il-86 was handicapped by the absence of a powerful, fuel-efficient turbofan engine comparable with those available to Western designers. The Il-86 uses four Kuznetsov NK-186 turbofan engines, each rated at 13,000 kg (28,660 lb) thrust, compared with the Tristar's three Rolls Royce RB 211-22B turbofans which in 1972 were each rated at 19,050 kg (42,000 lb) and the General Electric CF6-50C turbofans of 23,130 kg (50,000 lb) thrust, two of which power the 300-seat European Airbus A-300 which entered service in 1974.

In relation to its resources the Soviet Union almost certainly committed a tactical error when it decided to produce a supersonic transport comparable to the Anglo-French Concorde. Following agreements signed in November 1962, construction of Concorde by Aerospatiale and British Aircraft Corporation began in 1965. Two prototypes flew four years later and the production machines began commercial services simultaneously for British Airways and Air France on 21 January 1976. Power was provided in the early production models by four Bristol (Rolls Royce) Olympus 593 turbojet engines rated at 17,260 kg (38,050 lb) thrust with 17 per cent reheat. Later versions gave increased power and greater fuel economy at the designed cruising speed of Mach 2.0 to 2.2 at between 50,000 ft (15,250 m) and 65,000 ft (19,800 m).

A model of the Tupolev Tu-144 was shown at the Paris air show in 1965 and it was evident that the design was very similar to Concorde, though carrying slightly fewer passengers than the maximum of 144 for which Concorde was intended. The first flying prototype of the Tu-144 flew on 31 December 1968 so that it was the first supersonic transport aircraft to fly. On 26 May 1970 it became the first civil transport aircraft to exceed Mach 2.0. As testing proceeded, however, it became apparent that problems were being encountered, especially with the engines. The four Kuznetsov NK-144 turbofans were said to be failing to reach full power so that reheat had to be used constantly and fuel efficiency was low. Modified power plants failed to make the necessary improvements and after prolonged trials, including mail-carrying flights, the Tu-144 was quietly withdrawn from service.

Recent reports suggest that a new, much modified, version may be under development. This may be because the Soviet aero-engine industry has produced a more satisfactory large turbine in the new Kuznetsov engines used to power the Antonov

An-400 and the Il'yushin Il-96 (a derivative of the
Il-86 airbus). It is probably in the 20,000 kg
(44,000 lb) thrust class which would present
improved possibilities for a modified Tu-144.
Alternatively, the rumours may relate to a long-term
project for a new supersonic transport such as is
almost certain to be under consideration, as it is
in the West.

Organisational Problems
Aeroflot is subdivided into 30 directorates with
overall control residing in the Ministry of Civil
Aviation which was formed in 1964 from the former
Chief Directorate of the Civil Air Fleet[1]. Of the
present directorates, which are responsible for all
operational activities, 27 are, in effect, regional
airlines. At the peak of the pyramid, however, are
the Moscow Transport Directorate, which has
wide-ranging responsibilities including 'special
tasks', the Directorate of Civil Aviation of the
Central Regions and Arctic (Upravleniye grazhdanskoy
aviatsii tsentral'nykh rayonov i Arktiki) and the
Central Directorate of International Air
Communicaions (TsUMVS). This last was formed as a
subdivision of the Moscow Directorate in 1960,
became a separate Directorate in 1964 and was
renamed in 1971, using the name
Aeroflot-International Airlines as a trade name. The
Moscow Transport Directorate includes among its
tasks the operation of certain long-distance
services of national importance and the carriage of
the large numbers of passengers who travel to the
Black Sea resorts from the Moscow area, which is, of
course, the main area for the origination of
traffic. The heavy traffic within the most populated
regions of central European Russia, and services
between this region and the sparsely populated but
strategically important northern regions, are the
responsibility of the Central Regions and Arctic
Directorate. Its operations thus overlap those of
the Leningrad (formerly Northern) Directorate and
those of other Directorates which include Arctic
territories, rather as those of the Moscow Transport
Directorate overlap the flights run by the regional
directorates.
    The centralised administration, as well as the
uniform state ownership and combination of
international and internal services under the one
umbrella organisation, justify the description of
Aeroflot as the world's largest airline. Such a

massive structure must necessarily be divided into
functional units, but undoubtedly many of Aeroflot's
operational difficulties arise from the regional
overlap of responsibilities. It is, of course,
normal and inevitable that trunk, medium-range and
third-level air services overlap territorially and
that aircraft of many different operators and types
have to be handled on any substantial airport. What
is distinctive about the Soviet system is that all
the aircraft and civil airports are under the same
central management, but with regional authorities
which have their own plan to fulfil and norms to
which they must adhere. These are laid down by the
central command which may, nevertheless, insist on
priorities that make it very difficult for the
regional directorates to achieve their own given
objectives. Although they have limited command over
their own operations they do not have the power
conferred by independent ownership to argue with
their peers, and if they have recourse to appeal to
higher authority there is no separately constituted
body such as the U.S. Federal Aviation Authority or
the British Civil Aviation Authority to review
disputes; only the Ministry of Civil Aviation which
is, in effect, the pinnacle of command for the
organisation itself.

While there is no public discussion of the
wider issues of organisational problems (though
reforms have from time to time recognised the
out-of-date nature of the previous forms) outbursts
of annoyance and frustration among the various
authorities are commonly given space in the
specialist press. <u>Vozdushnyy transport</u> sent a
representative to investigate complaints that Moscow
Transport Directorate flights had a poor record for
timekeeping. It was argued that the directorate was
only partly responsible for its poor record because
70 per cent of flights were to or from other
directorates (<u>Voz. Trans.</u> 21 Aug. 1980).

The introduction of centralised computer
planning of flight schedules, which resulted in many
changes of timings, was blamed by some directorates
for increased difficulties experienced by crew and
passengers, resulting in falls in load factors. It
was claimed that services from Stavropol airport had
been subjected to more than 60 major changes within
two years (<u>Voz. Trans.</u> 28 Aug. 1980).

Bad planning of long-haul flights beyond the
control of regional directorates is also blamed for
congestion at provincial airports. Thus, the party
committee of Irkutsk airport complained of the

creation of an artificial peak between 10.00 and
11.00 hours, culminating in the schedule for six
large airliners, including two Tu-154s (a type
capable of being fitted-out for up to 167
passengers), being timed to take-off in quick
succession. As a result the airport might have to
handle five times its normal hourly passenger flow
(Voz. Trans. 7 Aug. 1980).
Central planning of postal and newspaper
flights also appears to conflict with regional
planning. Thus, daily flights from Khabarovsk to
Magadan were said to be carrying 20-30 passengers,
giving ample room for up to five tons of freight but
this was being sent on a separate postal/freight
service. In two weeks the Magadan aviapredpriyatiye
carried 124.5 tons of freight less than the plan
required, thereby wasting crew time, flying hours
and fuel, which, it was argued from Magadan, could
all be remedied by cancelling the postal flight
(Voz. Trans. 28 Aug. 1980).

Problems in Supply and Infrastructure
In addition to problems in design and production of
new types of aircraft, the endemic and perennial
weakness in relation to replacement of major
components and spare parts, which extends throughout
the Soviet economy, is clearly evident in the
aviation industry, despite its great national
importance. Possibly in some cases it is worsened by
this since there may well be direct competition for
equipment between civil and military needs. Such a
case could well be found in the supply of
replacement and reconditioned aero-engines. This
major, indeed vital, component, appears to be
responsible for keeping many aircraft grounded for
long periods.
Miniaviaprom is frequently criticised in
articles in Vozdushnyy transport for failures in the
supply of engines and spare parts. Examples are to
be found in the issues of 13 May, 12 June and
20 September 1980. In the last, it was stated that
shortage of replacement engines and spare parts had
kept 15 aircraft (9 Tu-154s and 6 Tu-134s) grounded
during the third quarter of the year in one
directorate.
Less sensational, but recognised as essential
for high productivity, the mechanisation of
passenger and freight handling and organisation of
airport services (including cleaning), provision of
containers and loading and handling equipment, are

all items that stimulate criticism. One complainant named Tyumen', Khar'kov and Sverdlovsk as important airports that lacked container-handling equipment (Voz. Trans. 7 Aug. 1980).

Runway construction and maintenance is a particularly important and demanding section of the infrastructure, especially in the climatic conditions that prevail in and after Russian winters. Pilots have complained that even at Moscow Domodedovo, principal airport for long-haul flights to the east, there is always one runway under repair, reconstruction having been 'going on for many years'. As a result, the same runway has to be used for take-off and landing operations, which was not intended by the airport designers. There is frequently a line of aircraft waiting to take-off with engines running, so wasting fuel as well as upsetting schedules. Frequently the repair operations on the runway lead to delays of 20-30 minutes (Voz. Trans. 4 and 9 Sept. 1980).

Delays in construction of new runways hold up the introduction of planned services. One pilot, in a letter to Vozdushnyy transport, asked the rhetorical question of why he kept seeing a covered Tu-154 (one of Aeroflot's most important aircraft) at Tyumen'. It was, he asserted, because although stationed there for regional services it had nowhere to go. Apart from Roshino, no other airport in Tyumen' oblast could take a Tu-154. Instructions had been issued for planning the extension of runways at Nadym, Nizhnevartovsk and Surgut but these could still take only the Tu-134 (which weighs little more than half as much). Furthermore, it was alleged that where a large aircraft, such as an Il'yushin Il-76, could be flown in, parking stands were lacking, e.g. at Nadym. At Nizhnevartovsk the lack of taxiways resulted in passengers having to be taken to the runway to board an aircraft. Freight ramps and container docks were also lacking at both these airports. Fuel supply was also a problem in this area, despite this being the country's main oil-producing region (Voz. Trans. 21 Sept. 1980).

Difficulty in phasing of improvements to enable best use to be made of equipment is illustrated by the long delays in overcoming problems at Yerevan and Kafan caused by the proximity of high mountain ranges. The importance of this route is shown by the fact that passes higher than 2000 m preclude effective road transport and the railway has to go a long way round through Azerbaydzhan. The introduction of the Yak-40, following runway construction which involved levelling and river diversion, resulted in a rapid rise of passengers to tens of thousands annually, requiring several flights daily. Owing to problems of mountain weather, however, the meteorologists advised that conditions were hazardous on 140 days per year. On such days special approach procedures were needed. Kafan was given priority in the allocation of the first SP-68 radiomayak (localising radio beacon) for special location on neighbouring cliffs. Proving this system would help in other mountain areas, such as the Pamirs. However, the plan was held up for want of a suitably equipped Yak-40, an An-24 which had the necessary fittings being not authorised for use in the mountain regions, so the scheme had to await provision of a jet 'flying laboratory' (Voz. Trans. 28 Aug. 1980).

Safety of Operations. In the above case it would appear that safety in the air is endangered by lack of equipment and delay in bringing into use devices already available because of inefficiency of supply. Another case in which equipment is partially responsible for less than optimal safety standards is in air traffic control. Gor'kiy was reported to be suffering from congestion, requiring controllers to accept twice or two-and-one-half times as many aircraft per hour as the twenty for which existing equipment was suitable. The controller allowed 2 minutes 55 seconds for each flight but at peak times he was overburdened and had to break his own rules, with obvious risks. The situation was exacerbated by shortage of personnel, including lack of deputies and understudies, so that controllers suffered from tiredness and could not take even a short break. Controllers on late or early turns often stayed overnight at the airport hotel but could not get supper because the restaurant closed too early (Voz. Trans. 30 Sept. 1980).

Although the Soviet Union belongs to the International Civil Aviation Organisation (ICAO) it does not publish accident statistics, as is desirable in the interests of safety. Accidents to Soviet aircraft are only known when they occur abroad or are witnessed by visitors, or involve travellers from the non-communist world. It is impossible to make satisfactory judgements, therefore, on the safety standards of Aeroflot or other Comecon airlines. Some generalisations are possible. The writer can say from personal

experience of extensive travel on these airlines
that he has not been concerned on many occasions
though he has seen badly worn tyres on aircraft in
operation such as would not be permitted on
safety-conscious airlines in the West. Air hostesses
are careless in checking seatbelts of passengers and
themselves commonly remain standing during take-off
when on major western airlines they would sit down
and strap themselves in. This is not unimportant
because if a stewardess is thrown off balance she
may cause injury to a passenger or to herself, in
which case she may not be able to carry out her
proper duties if an emergency develops. Male
stewards are not normally seen on Aeroflot flights.
    On the other hand, safety consciousness is
constantly demanded in the specialist press. All air
and ground crews are reminded of their
responsibilities in all their work, and crews and
administrators are apparently encouraged to write to
Vozdushnyy transport if they see failures to observe
the proper practices among their superiors or
elsewhere. The correspondent who wrote regarding the
delays in taking off from Domodedovo, for example,
also pointed out that these and other delays after
start-up of engines not only wasted fuel but
resulted in reduction of the safety margin if bad
weather required diversion to another airport at the
end of a long flight.

Conclusions
Too much should not be made of the complaints which
are published in the specialist press. These provide
a useful and possibly effective safety valve for
operatives and may give an excessive impression of
widespread inefficiency. There is no doubt that most
of the time Aeroflot flights operate with safety and
excellent timekeeping. Nevertheless, it is clear
that the vast organisation behind civil aviation in
the USSR requires constant vigilance on the part of
planners and directors and some bottlenecks and
problems occur that are not as easily solved as they
might be in advanced western countries.

[1] The most detailed account in English of
Aeroflot's evolution and activities is in Hugh
MacDonald's Aeroflot (1975). For a brief account of
history, services, and organisation, including a
full list of directorates, together with comparative
statistics for all transport modes, see Symons and

White (1975), Chapters 6 and 7.

REFERENCES

Alexander, Jean (1975), <u>Russian Aircraft Since 1940</u>,
     Putnam, London.
MacDonald, Hugh (1975), <u>Aeroflot: Soviet Air
     Transport Since 1923</u>, Putnam, London
Symons, L. and White, C. (1975), <u>Russian Transport,
     an Historical and Geographical Survey</u>, Bell
     and Hyman, London.
Symons, L. (1975), 'Soviet Civil Aviation -
     objectives and aircraft', in <u>Economic
     Development in the Soviet Union and Eastern
     Europe</u>, Vol. 1. ed. Z.W. Fallenbuchl, Praeger,
     New York.
Taylor, J.W.R. (ed), <u>Jane's All the World's
     Aircraft</u>, Macdonald and Jane's, London, annual.
United States Air Force (1978), <u>Soviet Aerospace
     Handbook</u>. A.F. Pamphlet 200-21, Dept. of the
     Air Force, Washington.

Serials

<u>Airports International</u>, London, monthly.
<u>Aviation Week and Space Technology</u>, New York,
     weekly.
<u>Flight International</u>, London, weekly.
<u>Grazhdanskaya aviatsiya</u>, Moscow, monthly.
<u>Interavia Aerospace Review</u>, Geneva, monthly.
<u>Jane's All the World's Aircraft</u>, Macdonald and
     Jane's, London, annual.
<u>Narodnoye khozyaystvo SSSR</u>, Moscow, annual.
<u>Transport i svyaz'</u>, Moscow, occasional.
<u>Vozdushnyy transport</u>, Moscow, three issues per week.

Table 5.1 Growth of Soviet Civil Air Transport

Soviet Air Transport

	1940	1965	1970	1975	1980	1982	Increase 1965-70	Increase 1970-75	Increase 1975-80	Increase 1980-82
Routes total (000 km)	146	481	773	827	996	1026	292	54	169	30
Average annual percentage increase	...	...	...	...	...	...	12.1	1.4	4.1	1.5
Routes within USSR (000 km)	144	435	596	645	780	811	161	49	135	31
Average annual percentage increase	...	...	...	...	...	...	7.4	1.6	4.2	2.0
Passenger traffic (milliard pass-km)	0.2	38.1	78.2	123	161	173	40.1	44.8	38	12
Average annual percentage increase	...	...	...	...	...	...	21.0	11.5	6.2	3.7
Passengers carried (million)	0.4	42.1	71.4	98.1	103.9	108.1	29.3	26.7	5.7	4.2
Average annual percentage increase	...	...	...	...	...	...	5.9	5.3	1.1	2.0
Freight (incl. mail) (million ton-km)	23.2	1338	1877	2590	3094	3030	539	713	504	-64
Average annual percentage increase	...	...	...	...	...	...	8.1	7.6	3.9	-1.0
Freight and mail carried (000 tons)	58.4	1228	1844	2472	2989	3105	616	628	517	116
Average annual percentage increase	...	...	...	...	...	...	10.0	6.8	3.5	1.9

Source : Narodnoye khozyaystvo SSSR, various dates.

Chapter 6

ROAD TRANSPORT AND THE SOVIET ECONOMY

Martin Crouch

In contrast to the railways, motor transport in the
Soviet Union plays a relatively minor part in the
national economy. Motor traffic, goods and passenger
combined, accounts for only 14 per cent (by ton-km
plus passenger-km) of the volume of rail transport.
None the less its importance, both relatively and
absolutely, is growing rapidly and it plays a vital
part in certain areas of the national economy,
notably - on the freight side - in the agricultural,
mining and construction industries, and - on the
passenger side - in the urban economy. It faces
real problems however and, despite its recent
expansion, has great difficulties in meeting the
demands made upon it, as the Soviet leadership has
publicly acknowledged. The Central Committee
Resolution of December 1982 on improving the
planning, organisation and efficiency of transport
enterprises is an example of the high priority now
being attached to transport, not least the road
sector, by the post-Brezhnev leadership.
    This chapter will consider these problems and
their overall significance, focusing first on
freight and then on passenger transport. In each
case problems will be discussed in two main
categories, inadequacies of provision and
inadequacies in planning and co-ordination. In the
conclusion it is hoped to suggest ways in which
these case studies illustrate wider aspects of the
Soviet political and economic process.

Freight Transport
In the USSR the transport of freight by motor
vehicle is always known as motor transport, never as
road transport. This is because, given the relative
roadlessness of much of the country, a lot of motor

transport does not take place on roads but on farm tracks, across open country and on winter ice and snow or in quarries and mines. No precise figures of road to off-road motor transport exist, but 'common-carrier' road freight transport represents less than one third of the national total, so off-road transport can be assumed to be significant overall as a proportion. What follows will focus mainly on road transport, leaving the phenomenon of off-road transport largely to one side.

Either way, motor transport is of considerable significance in certain areas of the economy. It is, for example, the most labour intensive form of transport. Out of more than 13 million people employed in all forms of transport, about 8.5 million work in motor transport (including here off-road and passenger transport). Much of the output of the consumer goods industries, one third of the country's grain, 40-45 million tons of sugar beet and many other farm products are transported annually by motor vehicle (Ekonomicheskaya gazeta 47, 1974). Inter-city freight transport by road is still in its infancy, due to bad roads, high costs and administrative problems. In 1975 inter-city main road transport constituted less than 1 per cent of total motor-borne freight tonnage, most of such freight being very local and short-haul in character. The average distance moved by motor transport in 1982 was 16.4 km per ton, compared with 930 km per ton for rail transport and 435 km per ton for inland waterway freight (Narodnoye khozyaystvo SSSR, 1982). So its significance is considerable, but generally local, and largely confined to certain sectors of the economy and acting as a feeder network to the railways. The major exception to this is in the Far East where there is often no railway to feed and the road may be the only available surface link. Road hauls can therefore be much longer in Siberia than in the western regions.

Equally striking meanwhile is the considerable growth in motor freight transport in recent years. In 1982 Soviet trucks carried 25.2 billion tons of freight, a three-fold increase since 1960, and more than a thirty-fold increase over the 1940 figures. As a percentage of rail traffic, too, the volume of road freight traffic has risen, so the increase is in both relative and absolute terms (see Table 6.1). Inter-city freight transport, for example, rose seven-fold between 1965 and 1978, and the current five-year plan indicates further expansion. General (i.e. common-carrier) truck traffic is scheduled to rise by 40 per cent between 1981 and 1985 (Ivanov, 1981).

Table 6.1 Motor traffic as a proportion of rail traffic (freight and passengers combined in the ratio one ton-km equals one passenger-km by rail and one ton-km equals six passenger-km by road)

Year	Rail	Motor	Motor as percentage of rail
1960	1675.1	108.7	6.5
1965	2151.8	163.2	7.6
1970	2760.1	254.6	9.2
1975	3549.0	388.6	10.9
1982	3812.4	532.8	14.0

Source : Narodnoye khozyaystvo SSSR (Moscow annually)

Not all growth is, of course, desirable, as both Chapters 2 and 9 demonstrate elsewhere in this volume in the context of rail traffic[1]. There are, however, some solid reasons for the growth of motor transport, these points notwithstanding. Motor transport is more flexible and over short distances can be much more economical (i.e. defining 'short' as anything up to 200 km). Soviet economists argue indeed that transferring another 100 million tons of freight from the railways to the roads would save up to 120 million rubles annually and as many as 68,000 railway wagons (Ekonomicheskaya gazeta 11, 1981). This is not being done in any comprehensive fashion largely due to the importance of val (i.e. ton-kilometres) as a measure of success of road as well as railway enterprises. No one wants short-haul work. The RSFSR Ministry of Motor Transport has suggested that haulage rates obviously need to be looked at and revised in order to encourage a more rapid transfer to road of short-haul traffic (Ekonomicheskaya gazeta 47, 1974). But this has not so far borne much fruit, particularly as rail transport is still generally perceived as cheaper. In 1975, including such costs as road charges and industrial sidings respectively, road transport was 27 times more expensive than rail per ton-kilometre. This is partly a comment on the average length of haul, partly on this deliberate pricing policy which has traditionally favoured railways (Hunter, 1957). So the expansion of road traffic is in fact all the more impressive given these obstacles.

Roads and Roadlessness. However, growth has brought attendant problems, not the least of which has been the fact that the expansion in truck output and road usage has far outstripped the road building programme. The volume of road freight has risen by 4,400 per cent in the past thirty years, but the length of hard surface roads by only 300 per cent (Izvestia, 10 October 1981). Between 1966 and 1980 the increase in production of motor vehicles was 224 per cent, but the increase in hard surface roads only 64 per cent. The discrepancy is not quite as great as these figures suggest. Many roads have gained a larger carrying capacity through widening, some of the vehicles produced are exported and much motor transport takes place off roads, as already mentioned. None the less, the density of traffic on roads either already too heavily used or otherwise inadequate has increased dramatically. To put it yet another way, the USSR now has, according to Soviet economists, 21 per cent of the world's industrial output but only 7 per cent of the world's 'top quality' roads and railways. Only one third of the road network is metalled (Voprosy ekonomiki 3, 1980). The roads are thus 'grossly overloaded' and 'falling to bits'. Their effectiveness has decreased pari passu. The average speed of road traffic is declining, and hold-ups and traffic jams - even in the Siberian tundra - are commonplace (Izvestia, 4 February 1982).

To help counter this, the current plan calls for a considerable increase in road-building. Between 1980 and 1985 an additional 80,000 km of hard roads are planned for completion including 11,500 km of top category motorways. But it is freely conceded that this is inadequate, that at least twice this amount is needed, and that even the current plan has often not been met, particularly in the non-Blackearth regions of the RSFSR (roads are generally better in the western Ukraine, the Baltic states and the Caucasus)(Pravda, 23 April 1980).

Meanwhile there are major problems arising from the prevalence of non-hard surface roads, and from the old issue of sheer roadlessness. Dirt roads or crushed-stone 'white roads' which turn to dust in the summer and mud in the spring and autumn are all too common. These factors, as Nove has written, have

a 'literally and figuratively incalculable effect'
on the economy and on national life (Nove, 1980).

The economic costs of bad roads and
roadlessness (bezdorozhye) are vast, particularly
for agricultural transport which is in any case a
vexed and complex area with special problems, and
far more highly seasonal in character even than in
countries with a less critical climate. This is not
surprising given that as many as 25 per cent of
farms in the RSFSR for example do not even have
normal road communications (Izvestia, 4 February
1982). In the Tatar ASSR less than one per cent of
farm roads are asphalted. Annual losses to
agriculture from bad roads alone are officially put
at 5 to 7 billion rubles and are rising. Up to 5 per
cent of the grain harvest, and 10 to 15 per cent of
the hay crop is lost each year because vehicles are
driven over sown areas to avoid impassable roads
(Pravda, 1 August 1979). In Saratov oblast alone,
for example, the losses are calculated at 50 million
rubles. The 'bad roads seasons', as they are widely
described, i.e. spring and autumn, render many
trucks useless. During such periods up to 60 per
cent of the available tractor fleet on farms can be
in use towing trucks or delivering freight in lieu
of trucks, with an obvious economic knock-on effect
(Izvestia, 10 October 1981). Bad or impassable roads
put the average truck out of action for no less than
40 days a year nationally. Because of the fear of
the bad-roads season, much of the sugar beet crop is
harvested prematurely, resulting in considerable
loss. More than 6 per cent of the fruit crop
nationally is lost because of bad roads.
Furthermore, it is thought that at least one third
of the truck fleet and large amounts of fuel could
be saved if there were an adequate national network
of paved roads. Transport costs over dirt roads can
be as much as 10 times higher than over modern paved
roads.

There are considerable regional variations in
the effect that all this has on agricultural
transport, as Table 6.2 indicates, with the RSFSR
and Central Asia much more severely circumscribed
than the Baltic republics. Estonia comes out
noticeably better, which is consistent with that
republic's generally much more productive
agricultural record. Average income per rural
household is higher in Estonia than in any other
republic (Kerblay, 1983b).

In fact, this plainly affects more than just the
agricultural sector, and has a widespread effect

Road Transport and the Soviet Economy

Table 6.2 Paved roads by Republic (1980) (kilometres of paved road per 1,000 hectares of developed agricultural land)

Estonia	16.50	Azerbaydzhan	4.84	Uzbekistan	2.31
Georgia	8.67	Ukraine	3.93	RSFSR	1.91
Latvia	6.76	Belorussia	3.82	Kirgizia	1.74
Lithuania	5.66	Moldavia	3.76	Kazakhstan	0.39
Armenia	5.46	Tadzhikistan	3.30	Turkmenistan	0.37

Source : Kerblay (1983a)

throughout the economy. To take just two examples; in 1980, work on a new gas preparation plant at Urengoy in the Tyumen' oil fields (a priority matter indeed) was held up totally for four months because the connecting roads had been of clay only and were totally washed away in the spring floods (Pravda, 21 January 1981). The service road built alongside the BAM might, if developed and improved, act not only as an invaluable permanent service road, thus cutting down delays on the BAM, but also as a means of opening up the surrounding territory. Instead, it is 'falling to bits' and, though carrying six times its intended traffic, it is officially due to be written off in 1985 when the BAM opens fully to regular traffic (Izvestia, 30 August 1983). More generally, it has been estimated that 1.5 per cent of the output of the entire consumer goods sector is lost through bad roads and associated delivery delays or irregularities.

Plainly these dilemmas are not entirely the fault of the regime's investment priorities. The weather is a considerable factor, forcing, for example, the winter closure of the Georgian Military Highway. One 60 km stretch of the Frunze-Osh road has 50 active avalanche corridors. Spring and summer bring their own problems too; mud and floods in the spring, heat and cracking or melting even of the most modern roads in summer. Winter is often the best time for motor traffic, for frozen rivers can be crossed, and the frost makes otherwise impassable off-road areas negotiable.

None the less, the problems are considerable, particularly for the rural economy, and contribute greatly to the difficulties of Soviet agriculture. The social and cultural consequences of bad roads and roadlessness - rural isolation, the flight from the land of the young and skilled - are enormous; and widely seen as very damaging. Indeed, a considerable school of literary writers, the so-called 'village prose' (derevenshchiki) movement, has been a direct result. Their concern with rural themes has not only produced some impressive writing, but has helped to shape a tangible national (particularly Russian) mood of worry and regret since the early 1970s, attracting much popular and critical acclaim. Writers such as Abramov, Belov, Shukshin, Rasputin and, not least, Solzhenitsyn, have all dealt powerfully with the · social and cultural costs of the flight from the land, the loss of youth and skills to the town, and the sheer isolation of rural life[2].

Even if road access were improved dramatically, of course, rural mobility would still be very limited. As late as 1975 only eight per cent of rural households owned a car. This point was very effectively conveyed by the well-known writer Vladimir Tendryakov in his 1956 story 'Potholes'. A truck driver, delivering smoked fish on a bad road between two villages, picks up a random load of paying passengers - a practice that is common, because it is often the only means of transportation for ordinary people in remote areas, but illegal. The truck tips over and a passenger is gravely injured. Another passenger, who happens to be the director of a machine tractor station (MTS) in the next village, organises a rescue attempt, and the driver and some of the passengers arduously carry the injured man there on an improvised litter. When they arrive it becomes evident that the injured man needs expert medical attention, which can only be found by hitching a tractor to a cart and taking him back to the first village. At this point, the MTS director ceases to be co-operative, and refuses to release a tractor on bureaucratic grounds. The injured passenger is placed on a horse cart which sets off in the rain. But it is too late. Because of the delay he dies on the way, a victim not just of bureaucratic intransigence, but of bad roads and inadequate rural transport. As Kerblay has sugested, however, there is a paradox at the heart of this, namely that the regime has traditionally never really favoured individual mobility. While it may want better roads for economic reasons, sociologically it is not interested in the development of any kind of 'car-owning democracy' (Kerblay, 1983a).

The inadequacy of the road system then is plain enough. This leads to a related inadequacy: that of the motor vehicle stock.

Road Transport and the Soviet Economy

The Vehicle Stock.   There were in 1975 4.2 million
trucks in the USSR, a 55 per cent increase over the
previous ten years. Furthermore, there was an
increase in the average size of trucks and their
productivity was higher. New vehicles were being
added at a rate of about 70,000 a year, a figure
which grew to over 80,000 by the early 1980s because
of the contribution of the new Kama truck factory.
In fact, production of motor vehicles generally has
increased strikingly. It took from 1924 to 1971 to
build the first million vehicles, but only another
five years to build the second million. Output of
trucks trebled during the eighth and ninth five-year
plans, 1966-75 (Narodnoye khozyaystvo SSSR). There
is now a wider variety of models and more large
tractor-trailer or road-train units of KamAZ and MAZ
types have been produced in a concerted drive to
boost labour productivity. Behind this growth lay an
expensive programme of building new works and of
reconstructing or enlarging existing plants. Both
the Moscow (ZIL) and Gor'kiy (GAZ) works were
modernised, as were the truck plants at Ul'yanovsk
(UAZ) and Kutaisy (KAZ), and in addition, the
biggest enterprise of all - the new Kama truck
factory (KamAZ) at Naberezhnyye Chelny - was opened.
     In some important respects, however, the
vehicle stock is far from adequate. Very large and
specialised vehicles are notable by their absence.
Standard 2.5-ton GAZ and 5-ton ZIL trucks of
antiquated design remain the most common types,
robustly constructed but heavy, underpowered and
hence relatively inefficient to operate. Vans and
light trucks, dump trucks, tankers,
pipe-transporters, cross-country vehicles of all
types, and refrigerated lorries are all in
'chronically short supply' (Izvestia, 18 July 1981).
Several illustrations of this might be in order.
Although seven different refrigerated truck models
are built, production of all models is well below
target at present. There is no large refrigerated
truck (over 5 tons) available at all. Gosplan, while
aware of the problem, appears undecided as between
road or rail refrigeration, but meanwhile much of
the fruit harvest, for example, is lost annually
because of this situation: 'at present in places
horses and carts are doing the job refrigerated
truck transport should be doing' (Izvestia, 19
February 1981). The shortage of small cross-country
vehicles can be gauged by the fact that the
four-wheel-drive LuAZ or Volynyanka, which is in
production (at the rate of 11,500 annually in 1980)

is in such high demand that people 'are even prepared to trade in a car for one'. The alternatives to this jeep are not always very attractive:

> The builders of the BAM often hear the roar of the powerful engine of a KrAZ truck or a mighty 'Ural' as it approaches through the taiga wilderness. But all it contains is a single passenger - a manager coming to a production conference perhaps, or the cashier with the wages. The need for such journeys constantly arises in this great project, but they can hardly be made on foot, and the vehicle fleet is composed almost entirely of heavy trucks. ... It is not very economical to use a 16 ton juggernaut to bring in a box of nails, but that is what happens. (Ekonomicheskaya gazeta 3, 1977, quoted in Parker, 1979)

Overall the problem, however, is not that there is a surfeit of large trucks; on the contrary, in 1981 it was estimated that optimal fleet distribution figures nationally required 30 per cent of the vehicle stock to be 'large trucks' (i.e. over 5 tons (compared with an actual figure of less than 10 per cent. Similarly, there was a shortage of small-capacity vehicles (under 2 tons), under 10 per cent instead of optimally 29 per cent. The surfeit was hence of 2-5 ton trucks: no less than 81 per cent of the 1981 vehicle stock was in this category, the optimal figure being 41 per cent (Aksenov, 1980).

Theoretically, this optimal fleet distribution pattern could save up to one million jobs, and increase productivity by up to 25 per cent. More rapid renewal of the vehicle stock, better spares facilities, and a large-scale switch to diesel engines have all been canvassed as well, as ways of increasing efficiency (Kommunist 15, 1983). Only 8 per cent of Soviet trucks in 1981 had diesel engines. Up to one third of current output, however, is now diesel engined. Meanwhile, 23 per cent of trucks are idle at any one time due to spares problems, and, as Kommunist put it:

> It is common knowledge that the country's vehicle fleet includes a large number of vehicles whose age has considerably exceeded that for writing off. They are kept in operation with expensive inefficient and

> frequent major repairs, the labour costs of
> which are two to three times greater than in
> the production of new vehicles ..... At the
> same time, spare parts are extravagantly
> expended on subsequent operation after vehicles
> have received major repairs. It is not
> surprising that there are never enough spare
> parts.

A register of vehicles due to be written off and a
ban on major repairs to such vehicles has been
called for, but not so far implemented.

In short, there is clearly substantial room for
improvement in the size, distribution and
availability of the vehicle stock, despite major
efforts in the 1970s. Both roads and vehicles are in
inadequate supply, given the pressures on this
sector of the economy. The more critical political
problem for the 1980s may well be the area to which
we now turn, namely not adequacy of provision as
such, but the extraordinary misuse and
inefficiencies in allocation which appear to bedevil
the road freight industry.

Administrative Issues and Politics.

> The transportation bottleneck Russians bewail
> so often is due not primarily to any lack of
> trucks, but rather to a stupefying misuse of
> trucks already on the road. A visitor to the
> Soviet Union sees hundreds of small,
> medium-sized vehicles lumbering around,
> coughing out thick black fumes and breaking
> down with stunning frequency. What makes this
> sight so remarkable is that nearly all the
> trucks are empty. (Parker, 1979, quoting H.
> Meyer in Fortune, November 1974)

This is an exaggeration, but only just.
Official surveys indicate that nearly half of all
road journeys by trucks are unproductive
dead-mileage runs (Planovoye khozyaystvo 6, 1978).
Every third truck is running empty for at least 100
km (Ekonomicheskaya gazeta 11, 1981). Admittedly,
this is not significantly worse than in the West,
but given the expectation of central planning, and
the pressures on the Soviet economy, such misuse of
existing resources constitutes a live political and
operating problem. This will now be considered under
two headings:

174

'bureaucratic pluralism'; and
productivity, particularly labour productivity.

Bureaucratic pluralism - The complex and divided
lines of responsibility in Soviet politics often
produce centralisation without coordination, in the
form of an administrative-political machine that is
far from efficient and not always as responsive at
middle and lower levels to central demands as is
sometimes imagined. This phenomenon, often dubbed
'bureaucratic pluralism', or in Soviet terms 'narrow
departmentalism', is certainly very marked in the
transport industry (Crouch, 1981). There is, for
example, no All-Union Ministry of Motor Transport,
although there are All-Union Ministries in
the case of the railways, air, and sea
transport. This reflects the traditional view of
motor transport as a primarily local and
intra-regional service, a view that perhaps becomes
less relevant as its national importance grows.
There are ministries of motor transport for the
republics, however, and the most important of them,
for the RSFSR, acts for the whole country in the
provision of certain specialised services
(Automobilnyy transport 1, 1976). There are some
five thousand local transport organisations (ATPs)
under the RSFSR Ministry of Motor Transport,
responsible for local truck, bus and taxi services.
    Meanwhile, many enterprises, ministries and
departments operate their own separate fleets, not
always confining their activities to their own
specialised needs, but often carrying the type of
general cargo which the ATPs exist to transport.
Indeed, about three quarters of the truck fleet is
owned by separate enterprises or by Gossnab, a vast
state supply organisation set up in the 1960s and
which owns a large fleet as a middleman
theoretically in order to guarantee more regular
deliveries. The Ministry of Motor Transport
frequently claims that it would be more efficient to
have all but the most specialised transport under
its own control because the productivity of its own
labour is one third higher, and average unit costs
one third lower (Pravda, 27 July 1979). The average
size of an ATP fleet is 120, compared with only 17
in works fleets, suggesting that ATPs indeed do have
ample scope for productivity reforms, even if these
might not be as easily achieved as is sometimes
implied. It is also argued by the Ministry of Motor
Transport that safety would be improved overall if
separate fleets disappeared, for its own safety

record is better, and its drivers better qualified and trained in the first place. This is not a negligible point, for road safety and driving standards in the Soviet Union are not high, and it is thought that at least 35,000 people are killed on the roads every year (Parker, 1979).

Individual enterprises and departments, however, value the flexibility and independence that owning their own vehicles gives them and in practice little has been done to alter the existing pattern, despite the possible waste involved, despite the Central Committee passing a resolution on the matter in 1982 deploring (in familiar tones) 'narrow departmentalism', and despite some successful examples of coordination. In Moscow more than half the city's truck pool has been coordinated into an apparently successful common-carrier network and in Leningrad there has been since 1977 a scheme to coordinate freight transport to and from the docks, which has cut time spent in processing ships, railway wagons and trucks by 15 to 20 per cent. In the Kuban since 1981 there has been a scheme to centrally coordinate the harvest truck fleet which has led to considerable savings (Ivanov, 1981). In Georgia there is now a single agency for 'Specialised Transport and Automated Transport Systems' such as cableways and pneumatic tubes. But these are isolated examples. Furthermore, there has also been very little evidence of rationalisation within the ATP system itself. Although many ATPs are very large, a lot are small, with less than ten vehicles apiece, some with very high unit costs. There is also a marked tendency, particularly in the mountain regions of Central Asia, to use scarce trucks rather than cheaper and much more suitable (though 'unprogressive') donkey carts (Pravda, 17 July 1983).

Meanwhile, each ministry has its own regulations and conditions, its own planning and pricing system, its own pay and bonus structure. Not just freight transport but the whole economy is affected. Roads, for example: the funding and administration of road building and maintenance is typically complex and fragmented, and has become more so since the original decision to localise road finances in 1958. Only 15 to 20 per cent comes from the central budget, resulting in what many see as an uncoordinated muddle (Izvestia, 4 February 1982). Vehicle repair stations, to take one other case, are under 40 different ministries, despite a 'chronic' shortage of skilled mechanics and an obvious need for rationalisation. If a departmental vehicle breaks down in a part of the country where the ministry owning it does not have a repair facility, it may have to be transported enormous distances by road or rail to one of its own garages, even though facilities belonging to another ministry are close at hand. In the RSFSR 12,000 trucks were out of action in 1982 through a shortage of tyres alone. Coordination failures indeed are legion. A noisy scandal erupted in 1982 over misappropriated funds which had been intended for a tractor engine repair plant which somehow never got built, though the various central ministries involved were all under the impression that it was not only built but in production (Pravda, 18 December 1982). As a consequence of all these problems there have been repeated calls for more coordination and the setting up for the first time of a proper All-Union Ministry of Vehicle Transport and Road Management (Kommunist 15, 1983). If nothing else, coordination might produce some facts, it is argued. At present there are, for example, no statistics totalling the length of roads by maximum permissible axle loads. The widespread use on poor roads of very heavy lorries is, however, often identified as a really major problem.

This is not to suggest that 'coordination' is either simple or necessarily effective. There are real limits to such activity, particularly in the agricultural economy and the optimum size for any economic or political unit is rarely a simple and obvious matter. Paradoxically, there are occasional complaints about too much coordination and centralisation, even in agricultural transport. For example, a scheme to centralise dairy and creamery facilities in Novosibirsk oblast in 1980-81 produced widespread trouble, because the new facilities were too far from the producers (Pravda, 26 May 1981). But overall, this fragmentation of resources or 'narrow departmentalism' costs the Soviet economy huge sums, and is widely acknowledged to do so.

Labour productivity - This is a related, but also distinct area, where evidence of slack is clear, again throughout the national economy as well as specifically in road transport. It is nationally a key political issue for the 1980s. In road transport, one particular aspect which has attracted attention is the cost of loading and unloading, which alone employs up to 10 million people, many

of them in industry (Trud, 19 January 1980). The
wider use of containers, trailers and mechanised
loading and unloading systems is now believed to be
essential and usage is slowly spreading, with marked
effect, as has been shown by experiments in several
areas, such as Tula, Orel', Bryansk, Kaluga and the
Oka transport region. But low productivity,
featherbedding, report padding, disciplinary
problems, theft and so on are also endemic, though
rarely quantified nationally. A 1979 survey which
dramatically highlighted the possible extent of such
difficulties revealed that, of thirty trucks and
buses stopped at random entering or leaving Moscow,
not one was doing what it should have been. Another
recent social survey found that only 16 per cent of
those questioned thought that petty theft at work
should be a punishable offence (Simis, 1982).

The Wider Context. The picture is not entirely
black, but even so, all these shortcomings both of
the road network and vehicle stock and of the manner
in which it is administered and coordinated, have
wide ramifications for the national economy. The
Brezhnev Food Programme and the huge investments in
agriculture in the 1970s are now critically
dependent for their long-term efficacy upon better
rural transport. Problems of 'bureaucratic
pluralism' not only throw into relief larger
questions about the relevance to the 1980s of the
antiquated Soviet ministerial system (the failure to
coordinate the truck programme and the road-building
programme being perhaps the best instance), but also
raise questions about the 'authority-building'
problem-solving ability of the Soviet leadership. A
significant discussion in 1982 about the entire
Soviet ministerial system focused attention on this
area (Kommunist, 18, 1982), with a strong call for a
radical overhaul and streamlining of the whole
system. As Breslauer has recently argued, the Soviet
leadership is in the business not just of power for
its own sake (through patronage), but also of
creating and maintaining legitimate authority
(through achievement). Road freight transport and
the roads problem provide a worrying syndrome of
both qualitative and quantitative problems, for
which there is no painless political solution, but
which cannot be ignored (Breslauer, 1982).

Passenger Transport

The position of road passenger transport both in Western literature on the USSR and also inside the USSR itself is a somewhat ambiguous one, and it is often neglected in considerations of transport and the national economy. This is, perhaps partly, because it is largely an urban phenomenon, and therefore has often been academically compartmentalised away under 'urban studies'; partly because in real life its position is politically still a genuinely ambiguous one. As one recent critic observed: 'the question of whether road passenger transport, particularly urban transport, is to be considered a part of local government services or of the national transport industry has yet to be resolved' [3]. None the less, it is worth examining if only because it does play an enormous part in the running of the economy, notably in getting people to and from work. By 1983 there were some 50 billion passenger trips annually by urban bus, tram, trolleybus and metro systems, not merely more than ten times as many as on the railways but, since about 1970, more passenger-kilometres too. It is thus the largest overall provider of passenger services; in 1980, public services by road, including trams, accounted for 47.4% of common-carrier passenger-kilometres (rail - 36.0%, air - 15.8%).

The sheer scale and rapid continuing growth of this sector is a reflection on urban development generally and on low car ownership rates, for most of this traffic is generated within urban areas. More than 64 per cent of Soviet people now live in urban areas, increasingly in very large cities, and with car ownership levels only barely equivalent to those of the United States 60 years ago, demand for urban public transport is vast. Some of this demand is met by trams and trolleybuses or, in the largest cities of all, by metros. But since the Second World War an increasing role has been played by the bus, which is cheaper and more flexible, and whose operating costs have risen far less in recent years than those of other modes. Trams and trolleybuses, which between them catered for nearly 80 per cent of all urban passenger-kilometres in 1950, now account for little more than 10 per cent of the national totals, as Table 6.3 indicates. Inter-urban bus travel, by contrast, is rather limited; though it has risen five-fold since 1960, most of it consists of very lengthy suburban trips and the average journey length is little more than 40 km. True

179

long-distance and inter-urban coaching is therefore very limited, though increasing. Inter-city services are most important in southern Siberia and Central Asia, though most intensively developed in areas with a relatively good road system such as the Baltic republics, the western Ukraine and the Caucasus. In the RSFSR as a whole, however, the average inter-urban journey length was 61.8 km in 1981[4]. Rural bus transport is also somewhat limited both in provision and in its total contribution to the overall figures, although as it is incorporated into the category 'suburban transport' for statistical purposes, it is not possible to be precise about this. As with rural freight transport, rural passenger services in remote areas are often provided by special-purpose cross-country vehicles. There is a limited amount of international coach transport, mainly to Poland, Czechoslovakia, Bulgaria, Hungary and Finland. One area of real growth has been in both school and enterprise personnel transport which has mushroomed, in common with the trend towards separate truck fleets maintained by enterprises and ministries, as described above. According to at least one estimate, 'private' buses are now more numerous than those belonging to the public operators (q.v. Table 6.3).

Notwithstanding what follows, Soviet public transport compares favourably with that of many other countries. It can, and often does, offer cheap, reliable and frequent services, even in the coldest of winters. But there are problems, analytically and politically very similar to the road freight industry, and with a real bearing on the national economy, which need examination.

Inadequacy of provision - This is a key problem nationally. The continued rapid urbanisation of the USSR and the consequent growth of cities and urban commuting would have placed strain on any system. But this, coupled with the relatively low priority accorded to road passenger transport politically, has meant that the situation has in some ways deteriorated markedly and measurably in recent years. The national daily average journey-to-work time has risen, and even in 1974 was 53 minutes and rising, as against a desired policy norm of 40 minutes (Pravda, 8 August 1974). In Moscow and Leningrad the journey-to-work times are often as long as 90 minutes. The overall economic effects of this, not to mention the human consequences of adding an hour or two to the working day, are clearly very large. Even in major and relatively favoured cities like Gor'kiy 'people are having to leave work one hour early in order to find a place on the hopelessly inadequate rush-hour services' (Pravda, 25 May 1976). Other major provincial centres such as Baku, Odessa, Kazan', Astrakhan, Orenburg, Vladivostok and Smolensk have all been publicly criticised in similar vein, though the picture is varied. Cities such as Moscow, Leningrad and Kiev, which get preferential treatment, undoubtedly have very impressive public transport systems by any standards.

But the problems exist and are not confined to urban areas. V. Perevedentsev, the sociologist, after spending a month in the country around Pskov in 1982, wrote of the 'appallingly unreliable' nature of rural bus services (Sovetskaya Rossiya, 25 August 1982). Only one third of those that he used left on time; there were frequent last-minute cancellations; people were often forced to wait a whole day to make important journeys, sometimes because vehicles had been hurriedly commandeered for other purposes by local officials (Pravda, 28 January 1982). Inadequate bus services and poor rail or air connections have hampered the progress of oil, gas and other construction projects in Siberia by disrupting the duty rosters. High labour turnover at the showpiece nuclear power station at Volgodonsk in Rostov oblast has been blamed on very bad commuter facilities between the town and the plant (Pravda, 12 July 1981). In more general and positive terms, the need for adequate bus services from small depressed towns to larger centres is crucial if employment opportunities are to be maximised, for many small towns have very limited employment opportunities[5].

Meanwhile, in the absence of adequate urban services, more car owners are commuting, with obvious 'bottlenecks' of another kind. Despite all the disincentives, car commuting, particularly in the summer, can be quicker and more convenient. In Estonia, 40 per cent of car owners commute to work by car, saving on average 30 minutes a day, and the percentage is rising (Voprosy ekonomiki 7, 1978). In Moscow there are already about 500,000 private cars, and traffic is increasing at 7-8 per cent per annum. As car ownership and usage grows, so do calls for central car parks and other facilities to make car travel less unattractive than it (deliberately) is at present (Literaturnaya gazeta, 11 October 1978).

Table 6.3   Urban   passenger-kilometres   by   mode (percentage)

Mode	1960	1970	1980
Public bus	39.1	41.2	23.3
Metro	8.6	8.2	4.8
Tram	24.0	12.9	5.1
Trolleybus	9.7	9.8	5.6
Private (i.e. ministerial, departmental, enterprise) bus	5.3	11.5	25.6
Private car	2.5	3.8	27.9
Other (e.g. taxi, boat, railway)	10.8	12.6	7.7

Source: I.M. Yakushkin, Passazhirskiye perevozki na metropolitenakh (Moscow 1982)

The question of modal split is worth examining briefly. Officially the authorities are committed to a substantial increase in electric transport, for environmental and efficiency reasons. This has borne only limited fruit, as Table 6.3 indicates. Bus usage has grown in both relative and absolute terms, while electric transport has not. Furthermore, the really remarkable recent growth rates are in the two sectors that are least acceptable, namely the private car and the 'private' bus. Nationally over one third of all urban passenger-kilometres are now accounted for by car, a dramatic growth from virtually nothing in little more than a decade. Not only does this conflict with declared long-term policy, but it is obviously wasteful of scarce resources.

These inadequacies ought not to be overestimated, but they exist and do not appear to be diminishing. This is partly because road passenger transport has remained a relatively low-priority area at a time when demand has mushroomed. It is also a comment on the relatively inflexible economic planning system. Take vehicle provision, for example: the output of buses has doubled since the early 1970s and is now running at

about 85,000 vehicles annually. Total vehicle stock
is about 350,000, having doubled in a decade. The
range is better and more varied than it was and
includes both Soviet and Hungarian-built stock. But
there is still a desperate need for more
high-capacity buses and for vehicles built
specifically to cope with very cold or very hot
circumstances or cross-country work. As with the
truck industry, the tendency has been to misapply
the virtues of standardisation at the cost of
variety. The same sort of difficulty in getting
innovation on the production side has been very
marked in the tramcar industry, a consistently
neglected field. Much tramcar production is carried
out as a sideline activity in what are basically
railway workshops, notably in Riga and Leningrad.
Not only should the output of tramcars be up to 100
per cent higher than the present annual total of
600, according to a recent Pravda discussion, but
the state of the track is 'disastrous' in Leningrad
and 'even worse' in almost every other city (Pravda,
3 March 1983).

Administrative Issues and Problems. The sheer
inadequacy of provision, though a problem, is in
part historical and certainly understandable given
the factors outlined above. As with freight
transport, the key political issue for the 1980s is
probably the need to make better use of existing
resources, for coordination problems are
considerable and, despite much talk over at least
the last 20 years, very little has in fact been
done.

Bureaucratic pluralism - On the surface, the road
passenger transport industry is run on clear
national lines. Wages and fares are set nationally
(though there are minor variations, particularly in
the Baltic republics). Fares, which are generally
flat fares, have remained unchanged since 1948. Just
beneath the surface, though, there is a familiar and
complex pattern of bureaucratic fragmentation. Urban
transport is under the direction of the town or city
soviet ispolkom, but is subdivided so that tram and
trolleybus operations are separate from bus
services, and metros are another matter still; each
is responsible to its appropriate hierarchical
superior, buses to the Ministry of Motor Transport,
trolleybuses and trams to the Ministry of Municipal

Services, metros to the Ministry of Railways. Spares for trams come, improbably enough, from the All-Union Farm Machinery Association or the State Committee for Material and Technical Supply, metal from the city planning authorities, paint from the regional authorities, ticket-issuing machines from the Ministry of Machine Building for Light Industry. Electricity is supplied by the local electricity undertaking, not self-generated: this is cheaper, but it does add another administrative layer. As Pravda has expressed it, with untypical understatement: 'our urban transport suffers from a lack of administrative coordination'. Even Gosplan has two separate departments studying urban transport (Pravda, 3 March 1983). This not only means that there is no single comprehensive view of urban public transport, no single national agency formally and exclusively responsible, but it also creates a lot of avoidable local day-to-day problems, because there is not even at the local level a unified public transport organisation (Crouch, 1979).

The consequences in terms of resource misuse are large, though difficult to quantify. Spares shortages, for example, mean that on any one day an average of 25 per cent of the national fleet is off the road; a comparable figure in Western Europe would be 10 to 15 per cent less. The failure to control school and enterprise 'private' buses - discussed above - is another example of bureaucratic inertia, and very wasteful of resources. The average works bus driver carries no more than 20-25 passengers twice a day, at an estimated unit cost three times that of the public services. Furthermore, it sucks in scarce labour into what really amounts to a sinecure (Crouch, 1981). In places, the labour shortage in this industry is as high as 25-30 per cent, with a very high annual labour turnover too.

Labour productivity - There are two aspects to this worth referring to. One is the endemic human problem of discipline: drunkenness, private enterprise operations and so on. Brezhnev himself lectured bus workers in 1976 on the need to work harder ('The welfare and good spirits of the Soviet people depend on you and your work. Remember this and try to do better. Show more initiative and emulate the leading workers') and there are frequent Press complaints about the pilfering of fares or missing buses 'diverted' by their drivers to more profitable routes, the cash then being pocketed. Annual losses through drivers pilfering or passengers avoiding payment is perceived as a serious and growing problem: Tbilisi has achieved notoriety in this respect. The second aspect is a structural one. Plan fulfilment is measured in passenger trips; to maximise these, managers frequently create artificially short routes requiring a lot of changes for passengers. The resulting growth figures look impressive, but the underlying reality is that the journey to work is often artifically complicated, with obvious consequences for the smooth running of the local economy (Lewis and Sternheimer, 1979). There is some talk about going for quality of service (i.e. speed, realiability) rather than crude <u>val</u> quantity measures (numbers of trips, mileage). But given the predominance of the quantity factor as a success criterion in the Soviet economy, it is difficult to see such changes being made in isolation from much more comprehensive economic reforms.

Conclusions

This discussion may sound unduly negative. It should be stressed that road transport in the Soviet Union copes fairly well, given the demands placed upon it, and the often inadequate resources provided. There <u>has</u> been significant expansion and some real improvements in productivity, but it is characterised, too, by poor administrative coordination and an apparent inability to make the best use of available resources. In this respect, road transport typifies much of the Soviet economy, and the deep-rooted and complex nature of its problems, as well as the apparent immobilism of the political leadership. What is now termed 'bureaucratic pluralism' or 'authority leakage' in Soviet politics is only what Herzen identified somewhat more eloquently in the 1860s: 'there was a gigantic machine with a vast quantity of wheels and little wheels, cogs, levers, transmitters and belts. But the tragedy consisted of the fact that the wheels were not connected to one another. Thus the machine spun around in meaningless circles'[6]. 'Incrementalism', or muddling-through, is not necessarily inappropriate, but there is real scope for improving the interconnections and limiting 'narrow departmentalism'. The failure to make any real headway in this respect despite some 20 years of debate is not a hopeful sign, particularly given the virtual end of Soviet economic growth and the attendant problems this poses politically.

One final, albeit ambiguous, aspect: the Soviet second (unofficial) economy, which has a real functional value, could not survive in its present form without the road sector, for it is very heavily dependent upon road transport. All the road bottlenecks which exist in the operating of the official economy do therefore, in the wider context, need to be set against the real contribution which it makes to the operation of the increasingly significant and often functional unofficial economy[7].

NOTES

[1] This point has been eloquently discussed by Aleksandr Solzhenitsyn in The Gulag Archipelago Part 3, Chapter 3 (London 1975) in the context of the White Sea Canal, constructed by slave labour in the 1930s for the greater glory of Stalin, but with no obvious economic utility. In 1966 Solzhenitsyn visited the canal and, in 8 hours, observed precisely two barges, each laden with identical materials, one proceeding north and one south. 'Economic value zero'.

[2] See for example F. Abramov's four-volume work Brothers and Sisters, V. Rasputin's short stories such as Money for Maria and Borrowed Time, V. Shukshin's Snowball Berry Red, V. Belov's Carpenters Tales, and A. Solzhenitsyn's Matryona's House. All dwell on the remoteness of rural life. One particularly bleak account which stresses the isolation and roadlessness of the village is A. Amalriks's Involuntary Journey to Siberia (Collins & Harvill, London 1970), the record of a dissident's internal exile.

[3] Yu. Lachinov in Ekonomika i organizatsiya promyshlennovo proizvodstva, 7, 1982, translated in the Current Digest of the Soviet Press Vol. 34, No. 42. This ambiguity is not, of course, confined to the Soviet Union.

[4] The categories inter-urban and suburban and urban seem to be somewhat flexible. In order to increase revenue, many operators resort to what Pravda has dubbed a 'subterfuge' and relabel urban services as suburban or inter-urban in order to collect larger (staged rather than flat) fares. See Crouch (1979) for a further discussion of this

point. The CPSU Central Committee adopted a resolution in 1976 attacking such 'unacceptable' plan fulfilment methods. However, fares, the traditional source of revenue for such enterprises, now bear only a nominal relationship to costs. At least 50 per cent of finance now comes from subsidy. The metros of Moscow, Leningrad and Kiev are about the only sectors of urban public transport making a profit (the metros of Tbilisi, Baku and Khar'kov lose money). So the temptation to· indulge in money-raising malpractices and lessen dependence on subsidy is high. See Kovrigin (Moscow, 1978) for further details and figures.

[5] See Soviet Geography: Review and Translation, Vol. 22, 1981, pp.419-428 for a discussion of this point.

[6] Quoted by S. Sternheimer in G. Smith (ed.) Public Policy and Administration in the Soviet Union, Praeger, New York, 1980.

[7] On these questions see e.g. A. Katsenelinboigen, Coloured Markets in the Soviet Union, Soviet Studies, Vol. 29, No. 1, pp.62-85, and G. Grossman, The Second Economy, Problems of Communism, September 1977, pp.25-40. On the political functionalism of corruption and the second economy, see K. Jowitt, Soviet Traditionalism: the political corruption of a Leninist regime, Soviet Studies, Vol. 35, No. 3, pp.275-97. Probably 25 per cent of household food consumption comes through the second economy.

REFERENCES

Aksenov, I. (1980) Edinaya transportnaya systema, Moscow
Breslauer, G. (1982) Khrushchev and Brezhnev as Leaders: Building Authority in Soviet Politics, George Allen & Unwin, London
Crouch, M. (1979), 'Problems of Soviet Urban Transport', Soviet Studies 31, 2, 1979, 231-56
Crouch, M. (1981) Transport Policy in Britain and the Soviet Union: A Political Paradox. Policy and Politics 9, 4, 439-54
Hunter, H (1957), Soviet Transportation Policy, Harvard University Press, Cambridge, Mass.
Ivanov, V.N. (1981) Avtomobilnyy transport: problemy, perspektivy, Moscow
Kerblay, B. (1983a) Un handicap de l'agriculture sovietique: l'état des routes rurales, Revue

d'Etudes Comparatives Est-Ouest, 14, 2, (June), pp.5-24

Kerblay, B. (1983b) Modern Soviet Society Methuen, London

Kovrigin, A.G. (1978) Finansy zheleznodorozhnogo transporta, Transport, Moscow

Nove, A. (1980) The Soviet Economic System, 2nd Edition, George Allen & Unwin, London

Parker, W.H. (1979), Motor Transport in the Soviet Union, Oxford, School of Geography Research Paper, 23

Parker, W.H. (1980) The Soviet Motor Industry, Soviet Studies, 32, 4, 515-41

Petrov, E.V. and Alekseeva, I.M. (1983) Statistika avtomobilnogo transporta, Moscow

Simis, K. (1982) Secrets of a Corrupt Society, Dent, London

Yakushkin, I.M. Passazhirskiye perevozki na metropolitenakh, Transport, Moscow

Chapter 7

THE STRATEGY OF INTENSIFICATION OF FREIGHT TRAFFIC
IN THE GDR

Johannes F. Tismer

I. The Intensification Argument

For the past several years the demands of party
leaders and government officials in the GDR for more
intensive development of the national economy have
become more and more strident. Intensification as a
strategy of economic policy is intended to induce
the ministries, trusts[1], and enterprises to
utilise available economic resources as economically
as possible. Since market forces are paralysed,
administrative pressure for greater efficiency in
domestic freight traffic is applied. It is said that
the quality of economic planning in general and of
transport planning in the different areas in
particular is to be improved in order to allow the
economy to move on to a higher level of development.
During the current five-year plan (1981-85), the
volume of services (turnover) in domestic freight
traffic is not to be increased at all, while at the
same time industrial production is expected to rise
by 28 per cent to 30 per cent. This is to be
achieved first of all for the purpose of reducing
energy consumed. During the 1981-85 period, domestic
transport carriers are obliged by plan to consume,
for instance, 25 per cent less diesel fuel and 22
per cent less fuel oil. The GDR Transport Minister
emphasised the intensification argument in a
statement as follows:

> The task affecting all sectors of the national
> economy - that of no longer permitting
> transport services to increase proportionate to
> industrial and agricultural production and
> construction but, on the contrary, realising
> increased commodity production in the national
> economy with the same or even reduced transport
> services - represents a powerful demand that

must lead to structural changes.   Die Aufgaben
des Verkehrswesens...   4/1982, pp.112f.)

II. Modified and New Planning Procedures
Modified and new procedures in planning freight
demand, transport freight rates and transport
investments aim at reducing specific transport
expenditures[2].

1. Planning of Transport Demand, including the Use
of Loading Capacity.   In this planning area,
stricter standards are being applied and new
procedures instituted. The following are to be
emphasised:
(a)   The planning authorities reduce the transport
needs submitted by the economic branches in order,
it is claimed, to make them more realistic.
(b)   Optimisation of transport processes is another
measure to lower transport expenditure by shortening
the distance over which freight is hauled.
Optimisation affects primarily shipments of bulk
commodities, such as coal, construction materials,
ores, wood and grain, as well as products of the
chemical, iron and steel, and food industries. In
the GDR these types of commodities account for about
80 per cent of all rail transport services
(ton-kilometres). In order to make transport users
more aware of the distance as a cost factor, they no
longer receive their transport indicators in tons
just to determine loading capacity, but also in tkm.
(c)   Because road transport requires relatively
large quantities of diesel fuel, administrative
regulations definitely favour rail transport. During
the 1981-85 plan period, 15 million tons of freight
are to be shifted from road to rail. Except for
goods being shipped across the border and special
shipments, road haulage will be used primarily for
local transport.
(d)   The central planners are pushing for more
intensive utilisation of works transport by
employing its capacity for general transport tasks
and giving the pooling of works transport under the
direction of a guiding enterprise even more
attention than before[3]. A new organisational form
of works transport is to be based on branch
specialisation, which is now being widely
propagated. The purpose is to create larger
organisational units which are as homogeneous as
possible and are similar in their structure to those
in public road haulage. This is intended to make it

more convenient for administrators coordinating operational planning and, above all, to include works transport in regional transport balances, for it is by this means that general transport tasks can also be assigned to works transport. The energy shortage has increased the pressures for a centralised control of all road transport. Such control should become more effective if the planned employment of vehicles is subjected to central management and can form the informational basis for measuring energy quotas, it is thought.

(e) Efforts are being made to improve the consistency of performance in domestic freight traffic. Therefore quarterly plans for goods transport are being introduced which are to be coordinated with the annual plans. On the other hand, the quarterly and annual transport plans have to be combined with the general plans of production and distribution of goods quarterly and annually. Continuous and substantial divergences between monthly and annual plan targets have occurred in the past within transport and production planning, and also between both planning spheres.

(f) Plan controls are being intensified. Trusts and enterprises which report a need for transport are now obliged to prove at the end of a planning period that the loading capacity and transport services they requested were in accordance with their actual needs.

2. Planning Transport Freight Rates. In order to coordinate physical indicators (tons, ton-kilometres) with indicators of value, price planning is used to supplement planning of transport demand and the use of loading capacity. The administrative measures mentioned above of reducing the demand of transport users for loading capacity and the volume of services by setting higher transport requirement norms are receiving additional support by raising the rates for domestic freight traffic by an average of 60 per cent (Leyendecker, 12/1981, p.409). This measure took effect on 1 January 1982. It is intended to help reduce transport demand indirectly. Furthermore, freight rates for the various carriers have now been changed to make rail transport still more preferable to road haulage and transport on inland waterways more preferable to rail transport. The wagon-load tariff, which encourages making the best possible use of loading capacity, now also offers increased

benefits to transport users. On the other hand, heavy fines for loading delays are intended to ensure that goods are loaded and unloaded punctually.

3. <u>Investment Planning</u>. The instrument of investment planning is employed to strengthen the transport system through expansion and rationalisation.

The balance of investment funds allocated to the various transport modes was radically altered in the GDR's five-year plan of 1981-85: 70 per cent of planned transport investments are to be allotted to rail and water transport, primarily to rail transport; its share of total transport investment will probably amount to 55-60 per cent, while before it was just over 40 per cent (Wagener 1979, p.72). Investments for equipment prevail over construction investments by a ratio of about 3 to 1 (<u>Statistisches Jahrbuch der DDR 1981</u>, p.86).

The redirection of transport investment is taking place at the sole expense of road transport in the GDR. Here investment is limited to the most essential measures in areas with new construction and to carrying out road maintenance.

The available investments are aimed at achieving the greatest rationalisation effects possible, especially for the electrification of the railways. This represents an area of special emphasis, particularly since electrification can help reduce consumption of kinds of energy which are in short supply, such as fuel and diesel oil. In 1985 electrified rail transport is expected to account for 30 per cent of all rail transport. In 1981 it amounted to 19.9 per cent (Scholz, 6/1981, p.199).

Great importance is attached to rationalisation in the area of trans-shipment, loading, unloading and storage processes. Here investment is supposed to lead to shorter turn-round times for the various means of transport. Rationalisation effects are being sought primarily in places where the relations between industry and transport are especially close, i.e. at large traffic nodes. These account for over 50 per cent of total goods turnover in the GDR and therefore represent a point of special emphasis for investment during the 1981-85 planning period.

III. Microeconomic Implications of the measures being taken to reduce Specific Transport Expenditures

1. Implications of Procedures in planning Transport Demand.
(a) Employment of stricter planning standards - The official justification given for raising the norms for using loading capacity is that the needs submitted by transport users are said to be exaggerated. It is certainly true that users usually request more freight cars than are absolutely necessary for a specific shipment, because they know in advance that the request will not be met in full. However, there are no objective criteria for judgement and, without such criteria, attempts to plan more realistically cannot succeed.

The practice of determining transport needs by branch of industry is hazardous. An aggregation of planning data by branches certainly facilitates accounting work, since averages can be determined more readily and also provide a better overall view. On the other hand, this approach creates serious problems for the ministries when they need to break down or individualise aggregated targets for loading capacity and volume of services with reference to the trusts and enterprises under them. Obviously, criteria for determining transport needs which are derived from average data and norms are not appropriate in planning what assets will be supplied to an economic unit. Usually those transport users considered of lesser importance for the national economy are affected, in contrast to enterprises which, for example, produce for export and therefore are of significance for earning foreign currency. Transport users whose demand is given less priority, however, experience restrictions on their transport dispositions. These in turn can lead to more or less serious difficulties in carrying out plan tasks and thereby endanger operational success.

(b) Optimisation of transport relations - Optimisation of transport relations is intended to reduce the weight of the distance factor in transporting goods, since the output measure ton-kilometre is a function of the distance over which a certain volume of goods is transported.

However, determining this transport performance-indicator of the ton-kilometre presents considerable difficulties. The producing enterprises have a fairly clear idea of the volume of goods that

193

will be shipped within a certain period. However, they are not aware of the shipping volume equivalent in ton-kilometres, neither in terms of physical units nor in terms of value. What enterprise knows its demand for ton-kilometres over a month, quarter, or year, and why should it pay special attention to this measure of performance when transport costs are not a significant factor anyway? Records of transport costs in trusts and enterprises are said to be incomplete (Tessmann, Vogel 11/1981, p.370). Furthermore, transport prices do not reflect real transport costs, either for the goods handled and transported or for the route selected. It is therefore not surprising when the cost of shipping goods does not cause transport users to economise on services.

The pressure being exerted to adjust demand for transport services to a specific energy quota means that, in the present situation, the political decision-makers have no choice other than to insist on choosing the shortest transport route.

> Transport customers receive their transport indicators in tons and ton-kilometres, i.e. in terms of transport volume and transport turnover. Thus they are faced with the task of avoiding long-distance transports if at all possible and delivering and obtaining their products while using a minimum of transport services (Trunte, Paetzold, 12/1981, p.403).

Of course, it is not in the interest of trusts and enterprises when business relations which were considered advantageous have to be given up; when receiving or shipping goods becomes more difficult for one enterprise or another because access points to the transport network (sidings) are not available; when the goods delivered - if at all possible in block trains - cause storage problems for the transport user which cannot be solved easily, etc.

When there are restrictions on delivery arrangements because of administrative optimisation measures, it can also be much more difficult for the producing enterprises to change suppliers if a contract partner unexpectedly cannot deliver as promised, no matter what the reason. A necessary increase in the quota of transport services originally approved is usually not possible because of existing capacity and energy shortages.

(c) Shifting of freight transport from road to rail and centralisation of works transport - Limiting long-distance movement of goods by truck and including works transport in the integrated planning of domestic freight transport will bring grave problems for many enterprises; the main reason for the growth in works transport has in fact been the frequent inadequacies in the transport services provided to transport users by the common carriers. These inadequacies have led many enterprises to maintain their own fleets of goods vehicles so that they, at least to a limited extent, did not have to rely solely on outside transport services if these were needed. Efforts to become more independent proved to be necessary for more successful production activities. As far as the enterprises are concerned, this `is the real significance of works transport.

Limiting the scope of road freight transport and forced and specialised pooling of works transport, however, increasingly impair its significance for industrial enterprises and organisations. They are deprived once more of the opportunity to choose transport modes on the basis of cost, security, frequency, speed, etc. Bridging transport shortages is made more burdensome and this, again, makes it more difficult for trusts and enterprises to fulfil production tasks.

(d) Imposition of shorter time limits for plan fulfilment - Both consignors and carriers suffer additional transport disadvantages due to the imposition of stricter timing on plan fulfilment. An exact plan coordination (e.g. every quarter) between transport users and transport enterprises presupposes that coordination of economic activity functions well at the macroeconomic level, and that regionally separated economic activities organised on the basis of the division of labour can be incorporated, if at all possible, without delay. An efficient transport system can make an important contribution to flexibility. However, the ineffectiveness of coordination which is characteristic of socialist planned economies - transport bottlenecks are only one of the forms it takes - makes it extremely difficult to meet the performance plan on time; the lack of coordination forces 'unplanned' reactions on the part of common carriers and firms, particularly those with continuing access to works transport. If under such

195

conditions the time framework for transport operations is too limiting, it can happen that imposition of stricter timing in plan fulfilment, which is intended to reduce transport needs, will fail to achieve the desired result due to objective constraints inherent in the economic system.

2. Implications of Transport Price Planning. Transport prices do not reflect the actual transport situation, i.e. do not indicate the real scarcity relations between supply and demand. This, by the way, is also true for transport systems functioning within market economies.

The tariff system for domestic freight transport in the GDR pursues two main objectives. One is to simplify calculations of industrial delivery prices. To cover transport costs, for certain assortments of industrial goods uniform additional charges are levied. Standardising rates within an area facilitates the calculation of average costs for industrial goods transported. In the interest of general price stability, transport prices are in principle inflexible and remain unchanged over very long periods of time, as the time span between the most recent (1982) and the previous tariff increase (1967) shows.

The second main objective of the freight rates set for domestic carriers is to make consignors more cost-conscious in their demand for transport services. The point of raising the tariffs for domestic freight and making the system more differentiated, besides adjustment of prices to costs, is to direct demand for transport so that, for example, long-distance goods transport is limited, shipping by rail and on inland waterways is preferred against road haulage, shipping capacity is better used, and goods trans-shipment is speeded up.

These two objectives can only with difficulty, if indeed at all, be reconciled with one another. Charging enterprises with an average branch-related share of transport costs and the lack of flexibility in freight rates must necessarily weaken cost-reducing incentives, such as avoiding long-distance transport or achieving better utilisation of loading capacity. Enterprises that cannot cover their production costs and are operating at a loss are subsidised, usually with profits earned by more successfully managed enterprises. Furthermore, inflexibility of transport

196

prices quickly leads again to increased discrepancies between indicators of physical performance and indicators of value, so that the latter lose their incentive function of economising resources. As a consequence, accounting degenerates completely. This also explains the relatively large increase in transport rates which recently took place after they had been unchanged for 15 years. However, the increase would have had to be considerably higher if it were to reflect actual supply and demand relationships on transport markets in the GDR. As a result, the central planning authorities had to resort to the expedient of rationing transport loading capacity, transport services and energy, and of imposing quotas. The economy's transport requirements have then to be adjusted to these limits.

Of course, the measures taken to plan transport demand render it more difficult for transport users to make managerial dispositions, and promote conflicts between those affected and those responsible for taking the political decisions. Thus the GDR's Transport Minister observed that raising the quality of transport planning and accounting would only be possible by struggling against routine and traditional ways of thinking and acting. On 1 January 1982 a State Transport Inspectorate was set up, equipped with broad powers and charged with supervising those who participate in transport processes, to ensure that the new plan directives are observed. It is remarkable that the measures taken are, in the Minister's view, improving the quality of transport planning.

3. Implications of investment planning. Policymakers in the GDR hope that the measures taken in the context of transport planning, including price planning, will lead in the fairly short term to a reduction of the economy's transport expenditures in general and of energy expenditures in particular.

By investment planning, such objectives can be successfully achieved at best only in the medium term, due to the longer gestation periods required by investment projects. It is possible that effects on capacity and rationalisation which are induced by investments contribute to a lessening of strain in the transport system, if its reliability is thus strengthened and traffic performance processes become more productive. This depends on the volume of the investments and on the choice of investment projects: capacity planning has a decisive influence on both of these.

Experience shows that investments are made where transport capacity bottlenecks have occurred or where this is an immediate threat. The implementation of investment projects arising from this is, however, only delayed because of the limited availability of material means, so that improvements in capacity and rationalisation do not keep pace with the national economy's growing demands on the transport system. Thus, a lessening of strain in the transport system can hardly be effected in this way.

Energy planning also has a decisive influence on investment activity in the transport system. The fact that transport investments are concentrated on the railways, including more rapid electrification, has its rationale in the need to conserve deficitary kinds of energy. The GDR has to concentrate on producing energy from domestic raw materials, such as the coal it can use in power plants, to produce traction current for electrified rail transport.

An advantage of electrifying the railway is that electrical traction has a higher conveying capacity than other forms of traction. This improves rolling stock utilisation, assuming that the necessary complementary investments with respect to laying additional track, constructing more stations, maintenance, communications, etc. are also made, though as a rule these are delayed. Coordination problems diminish possible rationalisation effects, such as gains in productivity. The discrepancy between available transport capacity and transport needs thus tends to increase during the period of time between the taking of the investment decision and the completion of a given investment project in its complexity; the only solution is to try to bridge the gap in some organisational manner, by mobilising 'hidden reserves', that is, through plan pressure.

Investment activities directed at trans-shipment and storage have been particularly neglected. Therefore they show a comparatively low level of technological development and productivity in the GDR. The services on feeder lines and those connected with trans-shipment and storage amount to 40 per cent of the so-called productive services provided in the national economy and employ 25-30 per cent of all production workers. This is said to raise the cost of goods handled in these processes by about 40-50 per cent (Hanisch, 2/1981, p.49).

However, an improvement in this situation, which has existed for a long time, cannot be brought about simply by transport investments, but rather requires industry - on whose rail facilities most trans-shipment, loading and unloading takes place - to make more financial means or resources in general available. In this field, the cooperation of transport enterprises on the one hand and industrial enterprises as transport users and receivers of transported goods on the other hand is described as very unsatisfactory. Industrial administrations and their subordinate economic units give priority to developing production equipment and neglect complementary investments for the transport, trans-shipment and storage of goods. Thus, capacity mismatches lessen the transport system's economic efficiency.

It has recently proved impossible, either through investments or through various forms of so-called 'socialist team-work' (works transport pools, cargo-handling pools), to speed up goods movement. On the contrary: the average wagon turn-round time increased from 3.85 days in 1978 to 4.03 days in 1981 (Statistisches Jahrbuch der DDR 1982, p.207).

There are no reasons to expect that, in the investment planning of GDR transport, sufficient capacity and rationalisation effects will be achieved in the medium term to relieve the transport situation. The quality of the transport system will suffer because concentration of transport investments on rail freight means that fewer choices of carriers which offer special transport services are available to consignors. This applies particularly to road haulage. The shares of domestic carriers in freight traffic turnover (tkm) are expected to develop by 1985, compared to 1980, as follows (Trunte, Paetzold, p.402):

	1980	1985 (Plan)
Railway	70.9	82.0
Shipping on inland waterways	2.7	3.0
Road transport	26.4	15.0

The substitution of rail for road transport is obvious, while shipping on inland waterways is only marginally affected. There are no plans to promote its development in the future. In the past decade

199

performance of shipping by water had a decline of 8
per cent, while that of the railway showed an
increase of 36 per cent and that of road freight an
increase of 58 per cent (in the case of public
transport) and of 88 per cent (in the case of works
transport) (Gross et al., 2/1982, p.45).

IV Macroeconomic Implications of the Measures being taken to reduce Specific Transport Expenditures

The policy of intensifying the use of domestic
freight transport causes difficulties for economic
units requiring transport services in the following
important ways:
- administrative limits to the provision of
  facilities and continued restricted supply of
  services
- introduction of quotas restricting energy
  supplies for domestic freight traffic
- further limitations on freedom of choice between
  means of transport
- binding transport users more closely to
  transport plan deadlines
- stricter plan controls covering the use of
  loading capacity and transport services according
  to need, and to enforce improved use of transport
  processes at the planned time
- centralised management of existing works
  transport
- limitations on cooperative agreements between
  enterprises

  Naturally these restrictions also have
macroeconomic implications. Depending on the
circumstances in which they are economically active,
transport users have differing preferences with
respect to the demand for transport services.
Experience shows that these differing needs simply
cannot be met by a rigid system of administering
freight transport. Such practices limit the
economy's ability to specialise, thus lessening the
efficiency of the division of labour. This in turn
is not without influence on macroeconomic
productivity.

1.   Handicaps   affecting   regional
specialisation. The energy-supply problem is
considered of fundamental economic importance, and
it is beginning to receive due consideration in the
perspective plans of regional development and

particularly in the general traffic plans of the
districts and large cities of the GDR. These plans
are supposed to accommodate the requirements of
socialist intensification and rationalisation. This
applies, for example, in relation to long-term
planning of the development of transport
requirements, transport infrastructure, transport
organisation and the modal split. Traffic planning
and regional planning, thus far isolated from each
other, are being brought into a closer relationship.
Thereby,greater significance will be attached to the
transport factor, with the expected effect of
reducing interregional transport relations between
economic enterprises. These relations are already
mainly limited to the railway network and its
junctions. This favours regional economic
concentration. On the other hand, a continuing
limitation on the development of goods
transportation by road and investment barriers to
road construction obstruct regional decentralisation
of the economy.

A reduction in the supply of transport services
by, among other things, cutting down on the
long-distance transport of goods means that, to a
certain extent, intra-regional production relations
replace production relations between regions.
Enterprises may thereby be limited to a certain
degree in specialising by choice of production,
supply and marketing locations. Limitations on the
economy's ability to specialise, however, have
negative effects on its productivity.

2. Branch-related handicaps of specialisation. The
range and variety of services offered by transport
enterprises are also of significance for the
specialisation of economic units as consignors
within branches. This depends on the ability of
transport enterprises to meet special requirements
for transport users with services which go far
beyond the actual transport operation. The transport
system in market economies both reacts to transport
needs and offers services, thus promoting branch
specialisation. In the GDR on the other hand, it is
mainly the case that transport users must adopt the
administratively planned service level of the
transport system. For the most part, consignors have
to content themselves with the services supplied by
the railway as the universal means of transport,and
have few possibilities to use other means of
transport offering a more differentiated range of

services. This may benefit centralised operational
transport planning, but it impairs the economy's
ability to achieve branch specialisation, and hence
its productivity.

In most countries, regardless of their economic
systems, it appears to be the fate of the transport
sector that it is not organised according to
economic criteria, but rather serves as an
instrument of various policies, whether these be
employment policy, regional policy, social policy,
energy policy or other orientations. Generally,
important economic implications of such policies are
either overlooked or are deliberately accepted. The
implications are of macroeconomic relevance, because
the transport system is very closely connected with
regionally distributed economic activities on the
basis of the division of labour.

The political leadership in the GDR judges the
success of the activities described here by the
results, for instance by how much transport and
energy expenditures are reduced. The micro- and
macroeconomic implications of these results are not
mentioned. The energy shortage undoubtedly forces
the GDR's political leadership to take action.
However, the serviceability of the transport system
may be so impaired by these energy-saving measures
that, in the long run, the disadvantages for certain
areas of goods production could be greater than any
advantages gained by saving energy. Such conflicts
between economic policy goals reduce the
productivity of the national economy. The size of
the gap between industrial production on the one
hand, and transport services on the other, is
unquantifiable. Failures in the supply of transport
services, unreliable transport services, a lowering
of the service level, etc. - these occurrences are
regarded by transport users as particularly
disturbing. If, due to the existence of transport
bottlenecks, materials are not delivered punctually
and the continuity of production is affected as a
consequence, this delays plan fulfilment under
certain circumstances, but does not render it
impossible. During the actual process of plan
fulfilment, this simply means that care must be
taken to see that those economic activities being
promoted with a view to branch- and/or
sector-oriented policy suffer least from transport
shortages; thus, enterprises which are regarded as
having a particular macroeconomic importance (for
example, enterprises producing goods for export or

for military purposes) are given priority in the transport services.

It is also extremely difficult to measure, in both the micro- and the macroeconomic context, the various forms of productivity loss resulting from inferior transport services. Measuring such loss becomes still more difficult, the greater the distortions are in the price-cost system of an economy. Such distortions chiefly arise where, as in the GDR, political factors have a strong influence on economic activity; thanks to economic and political control mechanisms, in market economies with democratic, parliamentary systems of government these distortions can more successfully be kept within limits than in socialist countries, where such control mechanisms are virtually nonexistent.

V Conclusions
By investment planning and production planning in the GDR, the capacity and energy available to the transport system, whose development is less favourably treated than that of some other economic sectors, are limited. In the context of transport planning, the economy's transport requirements then have to be adjusted to the capacity and energy limits. By this means, intensification as a strategy of economic policy attempts to solve shortage problems by rationing transport services, or by applying 'plan pressure', instead of substituting production of transport services for production of industrial goods by resorting to effective methods of economic coordination in order to achieve a more balanced development of economic sectors.

There are no indications, however, that policymakers in the GDR intend to decentralise economic decision-making structures and give enterprises more room for managerial dispositions (choice of mode, suppliers, etc.). The present form of central economic administration will remain unchanged, even though experience has shown (not only in socialist planned economies) that the ability of a national economy to coordinate and specialise its activities, and thereby its efficiency, is weakened, the more economic administration is practised and the more policymakers, government agencies, business trusts, industrial associations, etc., intervene in economic and transport processes. This happens because such influences usually violate the principles which must be followed for optimal allocation of national resources to ensure a balanced economic development. Administrative intervention causes distorting effects on the accounting system of the national economy. Distortions in the price-cost structure or the system of plan indicators have increased to a particularly threatening degree in socialist planned economies, where it is practically impossible to determine the scarcities of goods and services (including transport services) in the national economy. Among other things, this is reflected in the low level of economic effectiveness which characterises the goods transport system of the GDR and the Soviet Union.

The superior ability of a market economy to coordinate and specialise its activities, for example in the Federal Republic of Germany, doubtless has more favourable effects on the development of its goods transport system, but even here they are weakened by government regulation of goods transport and intervention by organised transport interests. Thus, in the final analysis, the difference between the goods transport systems in the market economy of the FRG and the centrally planned economy of the GDR is only a matter of degree, in both micro- and macroeconomic terms. Even though policymakers in the Federal Republic of Germany basically accept the regulatory principles of market competition and profit earning for the best possible allocation and distribution of national resources, they still do not apply these principles consistently. Instead, they constantly manipulate the transport system for various political purposes which continue to have questionable effects; as a result, transport services in the national economy become more expensive and distortions arise in market data. For this reason, in general transport policymakers should constantly review the measures taken in order to ensure that they interfere as little as possible with the accounting system of the national economy and that steps are taken to improve its efficiency when political intervention has led to undesirable developments in transport markets. If such corrections are not made, economic decisions will lack any reliable informational basis. This becomes especially clear if we look at administrative guidance in the field of transport in the GDR, with its exceptionally negative implications for the effectiveness of the transport system and the productivity of the national economy.

NOTES

[1]   Trust stands for the term 'Kombinat'. It is a modern type of production unit in the various sectors of the GDR economy consisting of several juridically independent enterprises and managed by a director-general responsible for concentrated and efficient use of all the resources at the disposal of the respective enterprise. There are two kinds of trusts, which are either attached directly to a ministry or to an association of nationally-owned enterprises.

[2]   Specific transport expenditure indicates how many resources are required (for instance, energy of special kinds) to achieve a certain transport performance (millions of net tkm).

[3]   Juridically independent enterprises join together as works transport pools for the purpose of jointly carrying out transport tasks under the direction of a guiding enterprise. For part of the works transport in some districts this organisational form of works transport has existed for a number of years. In 1980 there were about 1100 works transport pools.

Chapter 8

PROBLEMS AND POSSIBILITIES OF AN ALBANIAN-YUGOSLAV
RAIL LINK

Derek R. Hall

Political and cultural relations between Yugoslavia
and Albania have rarely been smooth. The current war
of verbal vitriol being waged between these two
Balkan neighbours focuses on the status of Yugoslav
Kosovo - a region redolent with nationalistic
symbolism for both Albanians and Serbs. In the
economic sphere, however, balanced relations
continue : balanced in the sense that trade is
bartered on an equal value basis[1]. Relatively,
however, while such activities represent about 20%
of Albania's foreign trade (totalling about $1
billion in 1982) they comprise less than 2% of that
of Yugoslavia (Abecor, 1983). Prior to the 1981
ethnic Albanian riots in Kosovo, relations between
the two countries appeared to be improving. A mutual
defence pact aimed against the Soviet Union was
signed in 1980, alongside confirmation of an
agreement to exchange electricity. Most
significantly for present purposes, agreements were
also entered into providing for the physical
connection of the two countries' rail systems at
Albania's northern border, south of the Montenegrin
capital of Titograd. In the context of Albania's
gradual rail extension programme (in the face of
hitherto poor transport links with its neighbours),
and its widening trade patterns (value of exports
increased by 33% over the 1976-80 period), this
appeared a natural, if belated development.
However, subsequent antipathy between the two
countries coupled with the relatively lower
significance of the link for Yugoslavia, and that
country's deep seated economic problems, have
resulted in hesitation and prevarication on the
Yugoslav side.

The degree to which Albania will modify its
policies after the imminent demise of the country's

only post-war leader - Enver Hoxha - is open to
conjecture. Partly in response to the problems
with Yugoslavia, other transport and communications
links are gradually being forged with the outside
world, links which could eventually prove to be more
fruitful than that provided by what is currently a
rudimentary rail system with a severely limited
capacity.

Political Background
Although· physically and culturally remote, and,
since 1961 ideologically isolated, Albania can no
longer be regarded as Europe's forgotten corner. In
recent years the country has been gradually
extending its trade and cultural links with the
outside world, and while still proclaiming
self-reliance, improved and implicitly non-
indigenous technology coupled with commensurate
improvements in infrastructure is required if the
country's economic development plans are to be
maintained. Outdated capital equipment was at least
partly responsible for a slow down in industrial
growth for 1982 to 4.7% compared with a planned
8.5%. The People's Socialist Republic, however,
retains few political friends[2], and apart from the
United Nations, refuses to participate in
multilateral organisations and agreements,
particularly where these are perceived as being the
pawns of hegemonistic superpowers. In rejecting a
Greek invitation to attend a conference discussing
the creation of a nuclear-free zone in the Balkans,
the Albanian leadership pointed to the fact that two
of the other five Balkan states were members of
NATO, two were in the Warsaw Pact, and Yugoslavia
was linked to both superpowers (Zanga, 1983a).
Indeed, Albania's political relations with its much
larger neighbour, into whom it narrowly escaped
being absorbed in 1948, have deteriorated
considerably over the period from 1981, since the
ethnic Albanian riots in Yugoslavia's Kosovo
region[3]. For their part, the Yugoslavs point out
that Albania was the only European country to
boycott the Helsinki Conference, and, by inference,
that the Albanian leadership cannot be trusted
(Zanga, 1983g). However, despite the verbal battle
being waged between the two over the status and
aspirations of the two million Kosovars (Zanga
1983f; Reiquam, 1983; Baskin, 1983; Artisien, 1984),
the planned connection of their two railway systems
is being brought falteringly closer. Such a link

appears to be a natural progression for the Albanians, both as a further step in the continued expansion of their rail system, and as a means of ameliorating the transport bottlenecks constraining trade with Yugoslavia and other countries accessible through the Yugoslav rail system. For their part, the Yugoslavs now appear less than enthusiastic to provide a further vehicle for the infiltration of Albanian ideas. Paradoxically, however, although less than 2% of Yugoslavia's foreign trade is undertaken with Albania, the country's heavy hard currency indebtedness to the west, and the consequent return to closer economic ties with the Soviet bloc (Antic, 1981), would also appear to encourage an increase in the barter arrangements that Albania favours[4].

Significance of the Albanian Rail System

The historical antecedents of an Albanian rail system were not conducive to optimism. Nineteenth-century attempts by land-locked Serbia to forge a railway line to the sea across Albanian soil were unsuccessful. An Austrian-built light railway system within Albania was quickly destroyed during World War I, and much the same fate was met by an Italian railway-building programme in the Second World War. According to official Albanian sources (Anon, 1982), at the accession of socialist power towards the end of the war, apart from Iceland, Albania was the only European country without a railway. All external trade had been undertaken by Italian shipping companies, although the country's principal port, Durrës, could only handle two small vessels at any one time and possessed only one hand-winch of 200 kg capacity. Post-war socialist Albania re-established a rail network, extending it, step by step, to an overall distance covered of some 400 km. The major routes lie along the country's coastal plains and through the central Shkumbini valley (an east-west alignment cutting across the strike of the mountains, and a route taken between the capitals of Rome and Byzantium by the ancient Via Egnatia - see O'Sullivan, 1972).

The significance given by the Albanians to their railway building programme is indicative of at least four of the country's distinctive characteristics. The first is the relatively low level of industrial and urban development, such that only recently has the elaboration of a rudimentary rail network been necessary to meet changing economic circumstances. In recent years, however, the movement of goods by rail has been increasing at a rate twice that by road : in 1970 24% of all internal freight movement was by rail; in 1975 the figure was 32%, in 1980 37%, and 1982 43%[5]. Albanian sources now claim that the average cost of transporting goods by rail is about one third of that by road. By 1985 it is planned that 47% of all goods movements will be by rail. This figure is to be achieved partly by the completion of lines to Vlorë and the Yugoslav border at Han i Hotit, increasing track length to 440 km. However, 80% of the increase will come from an intensification of the use of existing lines : the number of trains per day will be raised by 50%, the average weight of a train will increase from 800 to 950 tonnes, and the current average waggon turnround time of 2.4 days will be 'further reduced'. The building of more sidings and branches - particularly the provision of railhead facilities at important factories - has a high priority. It has been argued that already the linking of the cement factories at Fushë-Krujë and Elbasan to the railway network is saving annually 50-60 lorries, 300 tons of diesel fuel and more than 300 sets of tyres (Milo and Leka, 1984).

Secondly, Albania's difficult topographic conditions constrain the comprehensiveness of any internal transport system. Rail transport is, of course, particularly inhibited by changes in gradient. A third distinctive characteristic is a prohibition on the private ownership of motor vehicles. Fourthly, Albanian railway building and track maintenance has been characterised by the mustering of large numbers of young 'volunteer' construction workers operating in one-month shifts : e.g. 'volunteers' working on the 22 km Lac-Lezhë section of the Durrës-Shkodër railway line (Figure 1), numbered some 17,000 young people, mobilised into brigades of 30 strong[6].

Apart from its transport function, the railway building programme in this utilitarian society performs at least three important socio-political roles. Firstly, in a country fostering Europe's highest birth rate[7], claiming full employment, and obliging all able-bodied adults to work, it represents an important source of employment. Secondly, for a society centrally directing its construction works programme and labour migration, it draws together young women and men from different parts of Albania and deliberately encourages

Figure 8.1: Growth of the Albanian rail system and other transport links

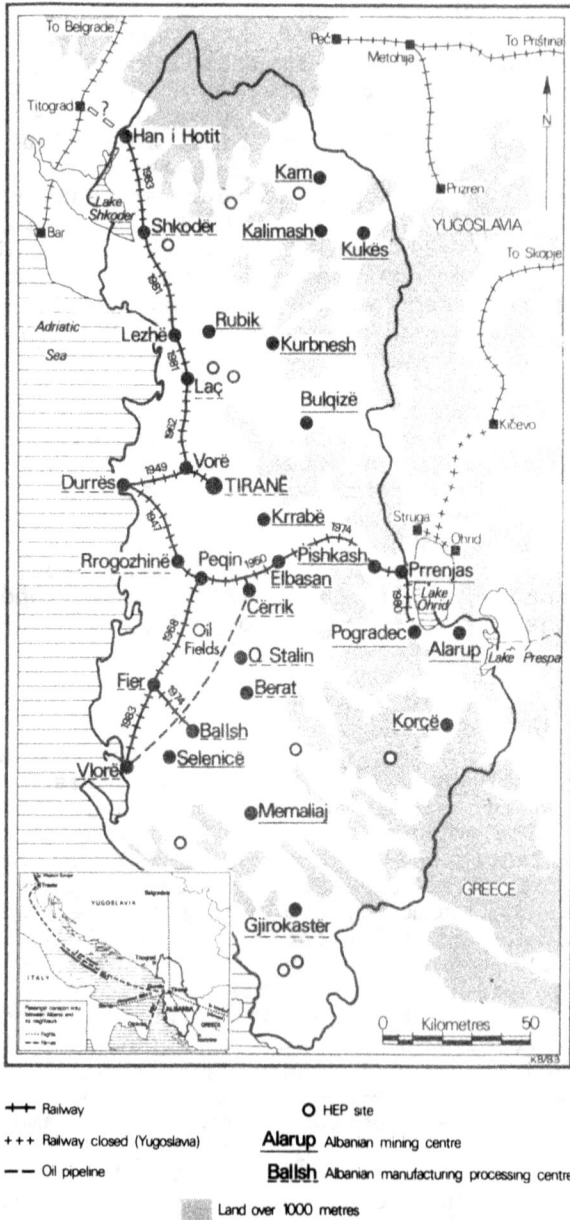

inter-regional marriages amongst them. This policy is aimed at overcoming long-held local and regional loyalties, thereby helping to eradicate traditional distinctions between the Ghegs of the north - generally mountain pastoralists - and Tosks - more settled southern valley dwellers. Officially these generic terms are no longer recognised: Albanians are now simply 'northerners' or 'southerners'. Further, the 'volunteer' youth workers are also employed to 'revolutionise' the citizens of the areas within which railway construction is being undertaken. Regular meetings are held with local inhabitants both to instil Marxist-Leninist fervour and again, to aid the intermixing of Albanians from different regional backgrounds and break down traditional social barriers.

Although most railway equipment has hitherto been supplied by other Eastern European countries - the current fleet of diesel-electric locomotives, for example, is Czechoslovakian in origin - 1982 saw workshops in Shkodër produce the first Albanian-built rolling stock (Anon, 1983a). Train movements meet with several constraints. Speed restrictions are common, apparently in response to a combination of relatively poor rails, sleepers and track bed, with frequent curves and gradients, particularly on the central line to Pogradec. Relatively infrequent passing loops appear to have been laid in this basically single-track system[8], and passenger train speeds average about 50 km/h (Muileman and Saltzman, 1981). Whether such low speeds represent long waiting times at passing loops is not clear. (No foreigner is officially allowed to use any form of Albanian public transport: such questions are, therefore, difficult to answer under existing circumstances.) The apparent relatively low use of the system at present might be thought of as reducing such waiting times for passenger trains, but again it is unclear as to whether, in this rapidly industrialising country, priority is given to passenger or freight movements : one would suspect the latter.

Axial routes out of the Durrës-Tiranë core area, and particularly from the premier seaport of Durrës itself, have been gradually extended to major industrial and raw material centres (copper and chromite mining areas in the mountainous north-east of the country being notable exceptions : Figure 1). Of prime importance since the mid-1970s has been the line to Elbasan, location of the country's (Chinese-built) metallurgical complex, and to the

sources of its raw materials at Pishkash, Prrenjas
and Pogradec. Again, since the mid-1970s the
country's oil producing area - between Cërrik, Fier
and Ballsh - has been well linked by rail. Thirdly,
the railway reached the superphosphate complex at
Lac in 1962. More recently, Shkodër has been
connected to the Durrës-Tiranë core by the northward
push of the railway to the Yugoslav border. This
northern city held the seat of the Roman Catholic
archbishopric, and for some time was Albania's
largest urban centre. But now the country is
officially atheist, and while historically national
leaders have tended to originate from northern
Albania, the post-war communist leadership has been
drawn predominantly from the south. Grounds for
suspicion that the north of the country and
particularly Shkodër have been discriminated against
in state policies since 1945 may be seen in the late
arrival of the railway[9] to this city with a
population comparable to that of Durrës. Finally,
but not least, the two most recent rail extensions
reached the country's second port, Vlorë (for a long
time linked to the Selenicë bitumen fields by a now
disused 31-km narrow-gauge railway line), and the
Yugoslav border at Han i Hotit/Bozau. This line is
intended to be linked up with the rest of the
European rail system via a new link to Titograd and
the Belgrade-Bar railway, opened by Tito in 1976
(Singleton and Wilson, 1977; Wilson, 1971). The
completion of this latter line doubtless gave the
Albanians the incentive to push northwards and
establish an eventual link.

An Albanian-Yugoslav Rail Connection
A rail link with Yugoslavia would, for the first
time, connect Albania's fixed-transport system with
the outside world to symbolise an overcoming of the
country's (always denied) isolation. There had been
possibilities during the early 1970s to take an
extension of the planned Elbasan to Pogradec line
round the north of Lake Ohrid to meet the Yugoslav
system near Struga. This option, was however, closed
when the Yugoslavs abandoned their 0.60 m
narrow-gauge railway south of Kičevo to Ohrid and
Struga (Figure 1). This line could have been
widened to standard gauge, but largely due to
the topography of the area, the line's running costs
had been estimated at five times the Yugoslav
average (Dokić, 1966, 222). This route is now served
by five buses a day (Jugoslovenske Želenice, 1981).

212

A key planned development in the Albanian system which would now facilitate ease of access to the Yugoslav system - were the Yugoslavs so willing - is an electrification programme. The construction of successively larger HEP stations in northern Albania has provided for surplus energy capacity and the export of electricity to neighbours. Whilst hydro-electricity can be notoriously seasonal - as the Albanians and Yugoslavs have recently found to their mutual cost after a period of prolonged drought[10] - the Albanians claim that several million leks could be saved annually by a switch from diesel to electric rail traction. While the country is self-sufficient in petroleum supplies, production is antiquated and faltering, and appears to have limited expansion possibilities as it suffers particularly from a lack of modern equipment. The Albanians cite three advantages in favour of electric traction : electric locomotives can pull heavier loads, especially on gradients above 1 in 80 (and stretches of the Elbasan-Pogradec line are inclined at between 1 in 40 and 1 in 25; they require only half the servicing time of diesel locomotives; and their maintenance costs are only between a third and a half of the latter (Milo and Leka, 1984).

Preparation for the cross-border rail link has been enthusiastically pursued by the Albanians (e.g. see Anon, 1980a,b, 1981 a-e, 1983b), despite worsening relations with Yugoslavia. Prior to both Tito's death and the ethnic Albanian riots in Kosovo, a bilateral agreement had been signed in April 1979. The rail connection was authorised in a protocol of April 1982, by which time the Albanian line had been extended as far north as Skodër. The protocol laid down that building work for the Skodër-Titograd link should begin on both sides in July 1982, and be completed in December 1983. Albanian construction work subsequently proceeded : April 1983 saw completion of 24 of Albania's allotted 50 km and 73 of its 106 major engineering projects (Zanga, 1983b). By August the Albanians were blasting rock around the villages of Hot, adjacent to the border, to provide 40,000 cu.m. of landfill for track bed around Lake Shkodër[11].

The Yugoslavs, on the other hand, had not even begun laying the track bed on their 25-km section. Montenegrin officials now argued that due to the great cost involved in engineering works around Lake Shkodër (including the reclamation of 12,000 ha of land[12]), Yugoslavia's economic problems would

delay participation in the scheme until at least
1985 (Zanga, 1983b). This was later confirmed by
Yugoslavia's director of the railway construction
directorate, who pointed to three problems. Firstly,
after the contract for the line was concluded
'prices changed'. Yugoslav legislation permitting
the railway to be built provided for 1,018 million
dinars at 1980 prices. A subsequent inter-republican
commission revised this estimate significantly
upwards to 2,100 million dinars at December 1982
prices, but the Yugoslav Assembly failed to approve
appropriate amendments to the original legislation.
The second problem facing the railway construction
directorate was that Yugoslavia's republics and
provinces had not been constant in meeting their
financial obligations for the project. By October
1983 only 377 million dinars out of a planned 850
million dinars had been received. Only Kosovo and
Bosnia-Herzegovina had completely fulfilled their
financial obligations, while Slovenia had
contributed none of its share. Lastly, a problem
over land expropriation had arisen (a constraint the
Albanians appear not to have to contend with). The
railway construction directorate and the 13th July
Agricultural Combine of Titograd could not agree
over the amount of the latter's land which needed to
be expropriated for the railway project and the
level of compensation to be paid for it[13].

Subsequently, the Albanian party daily Zeri i
Popullit (Voice of the People), has claimed that
Yugoslavia is deliberately attempting to isolate
Albania. Other interpretations have viewed such
tardiness as Yugoslav 'punishment' for Albanian
interference in internal Yugoslav affairs (in
Kosovo). This equally reflects the fact that the
railway represents a far greater importance for
Albania than for Yugoslavia, and that the latter
currently views the former as a rather irksome and
unnecessarily badly behaved child.

While not necessarily a reaction to this
situation, but reflecting a general requirement to
improve transport bottlenecks in foreign trade
relations, Albania signed an agreement with Italy in
October 1983 to establish a ferry service between
Durrës and Trieste for the transport of 'lorries,
goods and postal freight'[14]. If such a link were
merely to facilitate Italo-Albanian trade, several
ports along Italy's western Adriatic coast might
have been more appropriate. The choice of Trieste
suggests Albanian requirements for better access not
only to industrial northern Italy, but to central

and western Europe, effectively 'leapfrogging' potential Yugoslav impediments. Indeed, previously, the neighbouring Yugoslav port of Rijeka had been an important trans-shipment point for Albanian bulk cargoes. Prior to the 1981 Kosovo riots, nearly 250 cargo ships with Albanian products annually called at the port, while the Yugoslavs gave 'most-favoured-nation' treatment to Albanian lorries in transit, whereby fees were paid for up to only 500 lorries a year (Zanga, 1983c).

Albania's only post-war leader, Enver Hoxha, is 81 at .the time of writing. His eventual departure may herald a move away from Stalinist policies towards a greater flexibility in Albania's foreign economic and political relations as well as in its domestic policy. Foreign trade is generally expanding : that with CMEA member countries increased in volume by 20% in 1982 compared to the previous year[15]; in the west, trade with West Germany, which country, like the UK, has no diplomatic relations with Albania, increased by 35% over the same period[16]. But without improved relations with Yugoslavia, now soured over the Kosovo question, the railway link may yet prove to be unrealisable. Albanian-Yugoslav trade for 1983 and 1984 was planned at only $126 million annually, on an equally split barter basis, less than, for example, the amount that the Yugoslav republic of Slovenia earns from tourism. The 1983 first half-year trade figures for the two countries saw Yugoslav exports to Albania at $20 million, a third less than for the equivalent 1982 period, and Albanian exports to Yugoslavia at $28 million, a reduction of 25%. According to Yugoslav sources, late deliveries were common and were sign icantly slowing trade[17].

Any reappraisal of these relations now appears unlikely unless fundamental ideological changes take place in Albania. Ironically, such changes might drastically revise downwards the priority so far accorded to railway building and to rail transport there.

Summary and Conclusions

Several projects appear to emerge from the railway project:
1. Yugoslav hesitancy in actually completing the railway link due to: (a) poor relations with Albania and heightened suspicion of the country's motives[18]; (b) a low perceived priority -

while only 2% of total foreign trade is undertaken with Albania, the major item, hydro-electricity, already has its own infrastructure for exchange purposes; (c) the country's desperate economic situation, whereby the poorer republics (such as Montenegro) are particularly disadvantaged.

2.   If the railway is completed :   (a) continued bad relations could constrain its use;   (b)   the mineral resources of both countries are broadly comparable - chrome, zinc, copper, lead, nickel, lignite, poor iron ore, some petroleum - to further limit the potential of the railway for Yugoslav-Albanian relations;   (c)   the degree of spare capacity in the Yugoslav rail system may be insufficient to enable the Albanians to effectively use it to link with the rest of Europe.

3.   By forging ahead, Albania appears to recognise potential benefits of the link (but are these more apparent than real?) :   (a)   to   ameliorate bottlenecks in road transport links with Yugoslavia and Eastern Europe; (b) in the long term, just as Hamilton (1968, p.277) saw the Belgrade-Bar railway becoming a 'zonal multiplier' for the eastern two-fifths of Yugoslavia (given improved relations and a less impermeable boundary), similar economic stimulus could see Albania's industrial fulcrum elongating northwards along the railway to Shkodër.

4.   Ironically for Yugoslavia, with hard-currency problems and a desire not to be drawn too closely back into the Soviet bloc, increased barter arrangements with Albania could become an unavoidable expedient.

5.   As a second option, perhaps pointing to greater long-term pragmatism, the Trieste ferry link reflects both a (historically ironic) closer relationship with capitalist Italy, and an acknowledgement of a long-term relationship problem with Yugoslavia. Albania's February 1984 offer to the Yugoslavs of a wide-ranging cultural co-operation and exchange programme - with clauses patently unacceptable to them[19], has done nothing to ameliorate this position (Zanga, 1984).

NOTES

[1]   Indeed, Albania's constitution forbids the undertaking of any foreign credits or debts. Coming at the end of the partnership with China, some - notably Yugoslav - observers have seen this as Albania's means of repudiating and reneging on

all foreign debts. The Yugoslavs claim that Albania actually owes China at least $5.3 billion (American), the Soviet Union 120 million rubles, and Yugoslavia itself $42 million (SWB BBC Summaries of World Broadcasts) EE/7534/A2/1-2 7 Jan. 1984).

[2]   Recent improved relations with Vietnam have been notable as a possible reaction to signs of an apparent, if minor, Sino-Soviet rapprochement (Zanga, 1983d).

[3]   Within socialist Yugoslavia, per-capita income differences between the richest republic (Slovenia) and Kosovo, the poorest region, steadily widened from 1:3.3 in 1947 to 1:5 by 1980 (Singleton, 1983, 5).

[4]   Indeed, an agreement on goods to be exchanged during 1984 provided for an enlarged range of items and an increase of 12% over 1983 'in real terms' (SWB EE/W1259/A/3 20 Oct. 1983).

[5]   For the first six months of 1983 'rail transport' increased by 10% compared with the same period for 1982, while 'road transport in tons' increased by 8% (SWB EE/W1250/A/25 18 Aug. 1983).

[6]   SWB EE/W1134/A/16 14 May 1981.

[7]   An estimated 1980-85 annual population growth rate of 2.2% (Fincancioglu and Dinshaw, 1982).

[8]   While freight movements are more numerous, the frequency of Albanian passenger trains is currently unspectacular : six a day between Durrës and Tiranë, two between Durrës and Shkodër, Durrës and Fier (one extending to Ballsh), and Durrës to Elbasan (one extending to Pogradec) (Ward, 1983, 27; personal observation).

[9]   It was completed in November 1981, the Lezhë-Shkodër section having employed 7,000 'volunteers' (SWB EE/W1163/A/19 3 Dec. 1981). One could, however, equally counter-argue by pointing out that the southern city of Gjirokaster, birth place of the country's only post-war leader, Enver Hoxha, has yet to receive a rail connection.

[10]   For example, see SWB EE/W1269/A/19 5 Jan. 1984. During 1983 Albania only exported to Yugoslavia 35% of the planned one million MWh of electricity. Negotiations on 1984 contracts were subsequently held up, due, according to Yugoslav accounts, to Albania quoting 'very high prices'. (SWB EE/W1275/A/9 16 Feb. 1984).

[11]   SWB EE/W1252/A/15 1 Sep. 1983.

[12]   Earlier, publicity had been given to an Albanian-Montenegrin agreement to facilitate this (SWB EE/W1124/A/5 5 Mar. 1981).

Albanian-Yugoslav Rail Link

[13]   SWB EE/W1271/A/27 19 Jan. 1984.
[14]   SWB EE/W1261/A/18 3 Nov. 1983. This was
inaugurated two months later with a schedule
requiring the Venice-based Adriatic Navigation
Company's M.V. Tiepolo to call at Durrës on the 7th,
17th and 27th of each month (SWB EE/W1268/A/21 22
Dec. 1983). Later Rome radio reported that it was
the Italian government which had requested the ferry
link, together with the (re-)establishment of a
chair in Italian at Tiranë University and the
generation of more intense trade between Italy and
Albania (SWB EE/7597/A1/3 22 Mar. 1984). Following
the first Albanian trade delegation to the southern
Italian region of Puglia, the same radio station
stated that 'as early as next summer Otranto should
be linked with Durrës via a ferry and it cannot be
ruled out that the air route which linked Bari with
Tiranë may be restored' (SWB EE/W1274/A/2 9 Feb.
1984). The Yugoslavs appear to perceive this
Albanian opening up to Italy as a
(traditionally-seated) anti-Yugoslav policy (SWB
EE/7601/A2/1 26 Mar. 1984).
[15]   SWB EE/W1251/A/1 25 Aug. 1983. Albania
imports machinery, agricultural and transport
equipment, rolled ferrous and non-ferrous metals,
coke, products of the electrical engineering,
chemical and pharmaceutical industries, and exports
chrome, iron, nickel, bitumen, copper cables, food
and light industrial products. 1982 saw, for
example, the Romanians providing a lubricating-oil
plant and the Hungarians completing a
pharmaceuticals plant left unfinished by the Chinese
withdrawal in 1978.
[16]   SWB EE/W1259/A/2-3 20 Oct. 1983.
According to a Yugoslav report, over 40% of
Albania's trade was being undertaken with western
partners in the early 1980s (SWB EE/W1157/A/2 22
Oct. 1981), about half of that being with EEC
members.
[17]   SWB EE/W1254/A/3 15 Sep. 1983. Albania
receives mostly ferrous products, chemicals and
consumer goods such as refrigerators and television
sets from Yugoslavia.
[18]   Most recent Yugoslav accusations against
Albania of subversion and irredentism include the
official reception accorded in Tiranë to
representatives of what Belgrade sees are
anti-Yugoslav subversive emigré organisations, the
printing and issuing of postage stamps depicting
incorrect political boundaries, and the publication
of postcards of 'Albanian' towns which include

places in Kosovo (<u>SWB</u> EE/7675/A2/3 21 June 1984).
    [19]   For example, the document proposes that
'the exchanges interrupted by the Yugoslav side be
resumed in accordance with those agreements still in
force' (<u>SWB</u> EE/7561/A2/3 8 Feb. 1984).

ACKNOWLEDGEMENTS

Grateful thanks are due to Kevin Butler for the
cartography.

REFERENCES

Abecor (1983) <u>Albania</u> Barclays Bank, London
Anon (1980a) 'A new railway extending towards the
    north of Albania' <u>Albanian Foreign Trade</u>, <u>125</u>,
    5
    (1980b) 'The extension of the iron lines' <u>New</u>
    <u>Albania</u>, <u>1</u>, 26-7
    (1981a) 'A new railroad' <u>Albania today</u>, <u>58</u>,
    42-5
    (1981b) 'A new railway line' <u>New Albania</u>, <u>6</u>, 9
    (1981c) 'Inauguration of the Lezha-Shkodra
    railway' <u>Albania Today</u>, <u>61</u>, 57
    (1981d) 'The last kilometres ...' <u>New</u>
    <u>Albania</u>, <u>5</u>, 16-17
    (1981e) 'The newest railway of the youth' <u>New</u>
    <u>Albania</u>, <u>3</u>, 8
    (1982) <u>Portrait of Albania</u> 8 Nëntori, Tiranë
    (1983a) 'A collective of creators' <u>New Albania</u>,
    1, 13
    (1983b) 'On the new railway ...' <u>New Albania</u>,
    1, 12
    (1984) 'The newest railway of the country' <u>New</u>
    <u>Albania</u>, 1, 14-15
Antic, Z. (1981) 'Yugoslavia's trade with the CMEA
    countries increasing' <u>Radio Free Europe</u>
    <u>Research</u>, <u>6</u>, RAD/148, 19 May
Arnaoutovič, D. (1937) <u>Histoire des chemins de fer</u>
    <u>Yugoslaves 1825-1937</u>, Dunod, Paris
Artisien, P.F.R. (1984) 'A note on Kosovo and the
    future of Yugoslav-Albanian relations : a
    Balkan perspective' <u>Soviet Studies</u>, <u>36</u>, 267-76
Baskin, M. (1983) 'Crisis in Kosovo' <u>Problems of</u>
    <u>Communism</u>, <u>32</u>, 61-74.
Beaver, S.H. (1941) 'Railways in the Balkan
    peninsula' <u>Geog. Journ.</u>, <u>97</u>, 273-94
Dokić, M. (1966) 'Pruga Beograd-Bar : osnovne

postavke i efekti' Ekonomika Preduzeća, 14, 222-7

Fincancioglu, N. & Dinshaw, K. (1982) 'Fertility and family planning' People, 9, Special supp.

Hamilton, F.E.I. (1968) Yugoslavia : patterns of economic activity, Bell, London

Jugoslovenske Železnice (1981) Red voznje, Belgrade

Milo, K. & Leka, L. (1984) 'The development of rail transport' Albanian Life, 28, 19-20

Muileman, K.S. & Saltzman, M.L. (1981) Eurail guide, Pitman House, London, 11th ed.

O'Sullivan, F. (1972) The Egnatian Way, David and Charles, Newton Abbot

Reiquam, S. (1983)'Emigration and demography in Kosovo' Radio Free Europe Res. 8, RAD/186, 4 Aug.

Singleton, F. (1983) The regional diversity of Yugoslavia : economic, social and cultural aspects, British - Yugoslav Society Seminar Pap., Chelsea College, 26-27 Mar.

Ward, P. (1983) Albania, Oleander Press, Cambridge

Wilson, O. (1971) 'The Belgrade-Bar railroad : an essay in economic and political geography'. In G.W. Hoffman (ed.) Eastern Europe : essays in geographical problems, Methuen, London, pp.305-93

Zanga, L. (1983a) 'Albania rejects conference on nuclear free zone in the Balkans' Radio Free Europe Res., 8, RAD/150, 28 June

(1983b) 'Controversy over Yugoslav-Albanian rail link' Radio Free Europe Res., 8, RAD/93, 29 April

(1983c) 'Ferry connection between Italy and Albania set up' Radio Free Europe Res., 8, RAD/261, 15 Nov.

(1983d) 'Hanoi and Tirana on warmer friendship course' Radio Free Europe Res., 8, RAD/238, 10 Oct.

(1983e) 'The frigid nature of Yugoslav-Albanian relations' Radio Free Europe Res., 8, RAD/287, 30 Dec.

(1983f) 'The role of propaganda on the Albanian-Kosovar front', Radio Free Europe Res., 8, RAD/174, 28 Jul.

(1983g) 'Yugoslav-Albanian polemic continues', Radio Free Europe Res., 8, RAD/167, 13 Jul.

(1984) 'Tirana proposes cultural and other exchanges with Belgrade', Radio Free Europe Res., 9, RAD/22, 15 Feb.

Chapter 9

THE TRANSPORT SECTOR IN POLISH ECONOMIC PLANNING,
AND POLISH-SOVIET TRAFFIC CAPACITY PROBLEMS

Claes G. Alvstam,
Zygmunt Berman,
Andrew H. Dawson

> During the last years, the transport sector has
> become an even more obvious obstacle to the
> development of the country. It lacks ability to
> provide for the needs of economic life, and it
> lags behind the level of development reached in
> several other sectors of the economy.

> (Mieczysław Zajfryd, Minister
> of Communications, Nowe
> Drogi, April 1979, quoted in
> ZG 1979-04-29)

One of the chief aims of Polish governments since
the Second World War has been economic growth.
Emphasis has been placed upon the achievement of
ambitious industrial output targets and the
transformation of an economy which was characterised
by primitive forms of agriculture into an
industrial and urban one. Special attention has been
given, in line with the model of economic
development adopted in the Soviet Union, to the
expansion of those types of production which are
closely tied to localised sources of raw materials,
such as fuel and power, metals and chemicals. There
has also been some attempt - though to a much lesser
extent - to redress the regional imbalance in the
level of development between some parts of the
country which was inherited after the Second World
War, and factories have been built in what were
previously rural regions with surpluses of labour.
Some new investments, such as the opening up of
mineral resources in unindustrialised areas, have
served both policies, and all have given rise to new
demands upon the transport system, and especially
upon the Polish State Railways (PKP) - the national

railway system. Indeed, the relationship between economic achievement and the transport sector is fundamental. As Kusmierek (1980) has said 'either the goods trains are there and production can proceed or they are not and there is no production'. Accordingly, transport has been given a central role in the planning of the Polish space economy during some periods since 1945. Surprisingly, it appears to have been relegated to a much lower level of concern during others, and by the mid-1970s the lack of capacity had become a serious obstacle to further economic development. In this chapter we shall outline the development policies of the Polish authorities since 1945, and their implications for the rail network, before reviewing the achievements of the railways up to the 1970s. The demand for, and supply of, freight transport by the railways during the 1970s, and the reasons for the serious shortages which occurred, will then be examined.

Development Policy and the Railways

Immediately after the Second World War plans were drawn up for the spatial development of the economy. They were designed to deal with the severe problems of rural overpopulation and rapid population growth in the south-east of the country, and the shortage of alternative sources of employment to agriculture both there and in the north-east and north-west. Development of the Central Industrial Region - based on the Stalowa Wola steel works - which had begun before the war, was to be resumed, and major new centres of industry were to be established around Łomża and Ostrołęka and at Piła (in north-east and north-west Poland respectively - Figure 9.1). Other towns were also to be sites for industrial growth, and at the end of a 35-year programme there was to be a regional pattern of agricultural specialisation and a fully-fledged hierarchy of service centres across the country (Malisz 1974, pp.16-24). In other words, localised specialisation was to become common in all branches of the economy, and the need for exchange and movement between the different parts of Poland would accordingly be enhanced. Indeed, the National Spatial Plan of 1947 was built around the likely pattern of these movements (and thus followed a tradition in Polish planning which had been established by Chmielewski and Syrkus before 1939) with major developments of industry and services at the junctions or nodes in the movement pattern, and

222

Figure 9.1
Some elements of the Polish National Spatial Plan of
1947

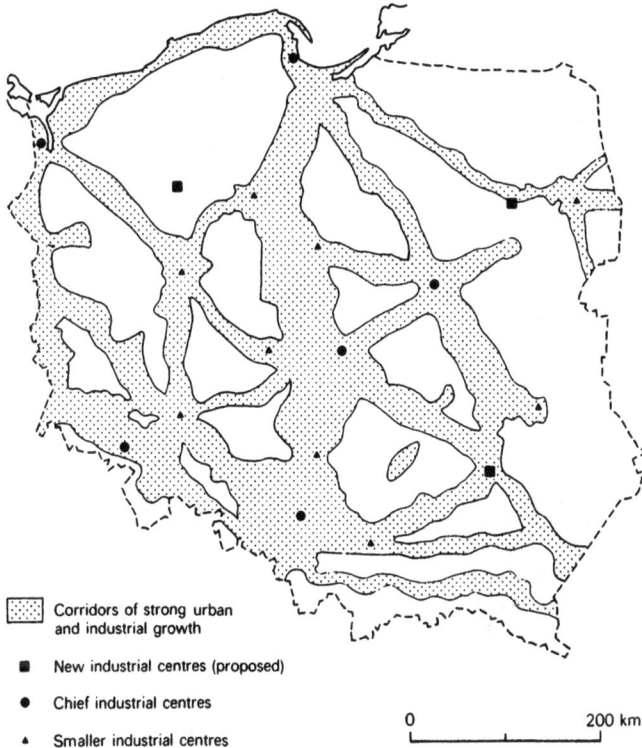

Corridors of strong urban
and industrial growth

■ New industrial centres (proposed)

● Chief industrial centres

▲ Smaller industrial centres

0   200 km

urban growth spreading out from those nodes along
bands or corridors containing the transport links
(Figure 9.1). Although some new transport facilities
were planned, it was the inherited pattern of the
railways which determined the detailed distribution
of both the nodes and corridors.

It was envisaged that a large part of the
movement of traffic would be associated with foreign
trade. Between the wars Poland's most important
trading partners had been in central, northern and
western Europe, and in 1938 eighty per cent of
foreign trade had gone by sea. Trade with the USSR
had been negligible. It was expected that a similar,
though not identical, pattern would develop again
after the war. In particular coal exports and
imports of iron ore from Sweden were resumed.
However, the displacement to the west of Poland's
borders had added the port complex of

223

The Transport Sector in Polish Economic Planning

Szczecin-Świnoujście to the country (as well as Gdańsk), and consideration was given to the construction of a steel mill on the western edge of Upper Silesia, close to deposits of metallurgical coal, which could be supplied with Swedish ore via the river Oder and the railways. The barges and ore wagons, which would otherwise have returned empty, could then have been used to take coal to Szczecin for export (Berezowski 1973, p.234).

In the event, none of these plans were accepted. Criticism of their authors during the onset of the Cold War led to their abandonment and the closure of the Central Planning Office which had produced some of them, and cleared the way for the adoption of economic development policies which had long been pursued in the Soviet Union, and in which the spatial element was largely absent, and for a reorientation of Polish trade links towards the east. From the late 1940s onwards plans gave the highest priority to the growth of output, especially in the producer-goods industries, and in so far as they were concerned with the distribution of industry, the plans took the simplistic view that it should be spread evenly or uniformly over the country. A highly ambitious six-year plan (1950-55) was approved which included the construction of a thousand new mines and factories, many of which were to be in rural areas. In fact, the plan proved to be over-ambitious. More than half of the factories were cancelled, and others took longer to complete than had been envisaged (Wróbel and Zawadzki 1966, pp.435-6). Nevertheless, the output of existing installations was increased considerably, new mines were sunk around the edge of the Upper Silesian coalfield, brown coal and copper deposits were exploited in central Poland and Lower Silesia, and new steel mills were erected at Nowa Huta, to the east of Upper Silesia, at Częstochowa and in Warsaw. There was a massive increase in the quantity of raw materials to be delivered to factories, of products to be distributed across the country or exported, chiefly to the Soviet Union, and of building materials required on industrial and urban construction sites. Moreover, because the increase in the industrial workforce exceeded the provision of new housing in the towns, long-distance commuting by rail became widespread.

Despite the restoration of Gumułka to power in 1956, economic policy during the late 1950s and 1960s showed little change in so far as transport was concerned. Many of the factories which had been begun under the six-year plan were not completed until the late 1950s, and although a New Location Policy was announced, it was only somewhat more restrictive than that which it replaced. Under the Policy industrial growth was to be concentrated in areas in which investment would give the best return in terms of increased output or where it would open up localised raw materials or make use of supplies of labour (Wróbel and Zawadzki 1966, pp.439-40). In the event much of the growth was achieved through the enlargement of existing plants, and large-scale commuting into the towns to work continued. Nor could major changes occur in the pattern of investment in industry, and in the consequent demand for rail services, immediately after the fall of Gumułka in 1970, though new aims and methods of planning were emerging about that time. Thus, the entire period between the late 1940s and the 1970s was marked by an emphasis on the fulfilling of industrial production targets, and particularly those of fuels, ores, building materials and other basic, but bulky, materials which were dependent upon the railways for their distribution. For example, the output of bituminous coal rose from 78 million tons in 1950 to more than 200 in 1979, that of coke from 6 to 20, and that of cement from 2.5 also to 20 (R S 1983, p. xxxiv).

These dramatic changes in the size and structure of the Polish economy, combined with a great shift in the direction of trade towards Eastern Europe, and especially the USSR, were reflected in the development of the rail system in several ways. Firstly, whereas most effort between the two World Wars had been put into building the 'coal line' from Upper Silesia to Gdynia, and to the connection of the Bydgoszcz and Poznań areas of what had been Prussia with the network in the area which had belonged to the Russian Empire, most of the new lines since the Second World War have been designed to facilitate movement between Poland and the Soviet Union and through traffic from East Germany to the USSR. In particular, connections were built betwen Tomaszów Mazowiecki and Radom (87 km) in 1949 and between Skierniewice and Łuków (161 km) during the early 1950s (Figure 9.2). As a result of this construction the length of normal-gauge track in Poland increased, at a time when many West European systems were being pruned down, from 22,500 kilometres in 1950 to 23,300 in 1970. Secondly, the capacity of the system was increased in the period up to 1970 through the electrification of about a

Figure 9.2
The railway network of Poland in 1980

Normal and broad-gauge Railway Lines in Poland

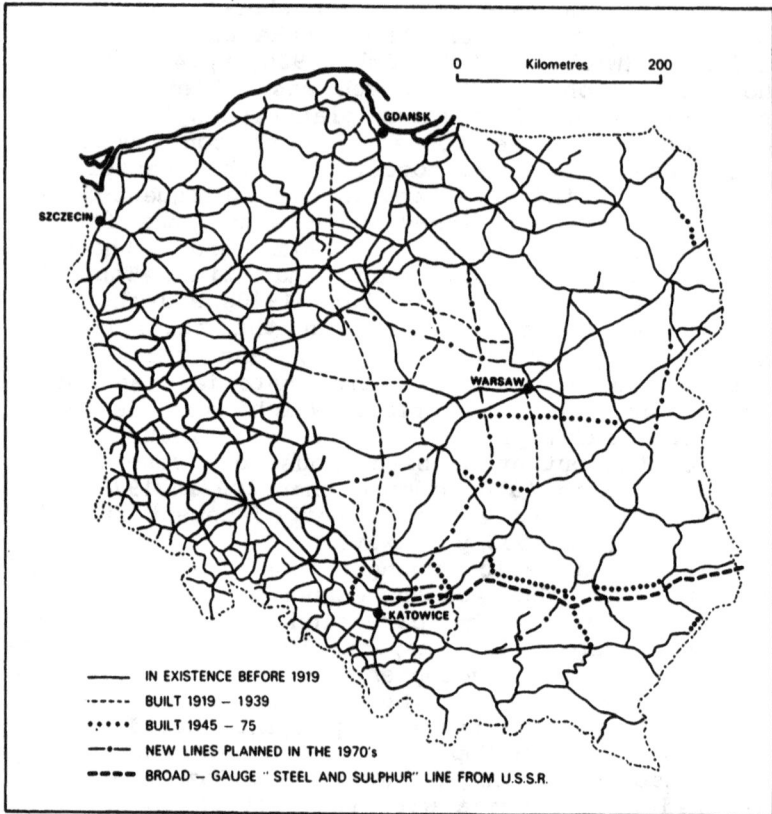

sixth of the track, chiefly on the most heavily-used routes in southern and central Poland; and all this allowed the quantities of freight and passengers which were carried to be increased very greatly. In 1950 freight carriers of all types in Poland handled 351 million tons, but by 1970 that figure had increased fourfold, and by 1980 eightfold. Moreover, despite the growth of road and pipeline transport the proportion of traffic on the railways had only fallen from 95 to 80 per cent in 1970 and to 68 per cent in 1980[1] (Table 9.1). Thus the Polish railways, which had carried less

than those of either France or West Germany in 1950,
were carrying more than the combined total of the
French, West German and Austrian systems by 1980.
We may conclude that a great increase in the demand
for transport services had been met largely by the
intensification of the use of the available rail
network.

By the 1980s freight traffic on the railways
was dominated by coal and coke, which accounted for
about half of the total ton-kilometres carried
(Table 9.2). Stone, sand, gravel and cement, metals
and metal products and ores together contributed
about a quarter. Almost all the coal and coke and
much of the metal and metal products originated from
Upper Silesia, and much of the stone, sand, gravel
and cement also came from there, from the Kielce
area and from the Sudeten Mountains. More than
three-quarters of all freight was sent to
destinations within Poland, and about half of the
rest passed over the borders, either on leaving or
entering the country, by rail. The remainder was
delivered to, or distributed from, the ports' (Głowny
Urzad Statystyczny 1981). Thus, the chief flows of
freight by rail were either local, such as of sand
from areas around the Upper Silesian coalfield to
backfill the mines there, or were of coal from Upper
Silesia to all parts of the country, to the USSR and
to the ports, or of iron ore from the USSR to iron
and steel mills in Kraków and Upper Silesia (Figures
9.3 and 9.4). Although Katowice voivodship covers
only two per cent of Poland, it accounted for about
half of all the goods loaded on to trains in the
country.

Thus, there were two very different periods of
economic policy in Poland between 1945 and the 1970s
in so far as the spatial pattern of development was
concerned, of which only the second was sufficiently
lengthy to allow its impact to be felt upon the
demand for rail transport and to enable the railway
system to be adapted to meet that demand. However,
early in the 1970s the authorities became aware that
the transport sector was no longer able to satisfy
the continuously growing demands upon it, and
several major projects were initiated with a view to
increasing the capacity of the rail, road and inland
waterway networks. For reasons which will be
described below these efforts did not succeed, and
by the mid-1970s the shortage of capacity had become
a serious obstacle to further economic growth.
Indeed, it was a transport crisis during the second
half of the 1970s which was one of the chief causes

Figure 9.3
The main movements of coal from Katowice voivodship
for domestic consumption in other voivodships and
for export in 1977.

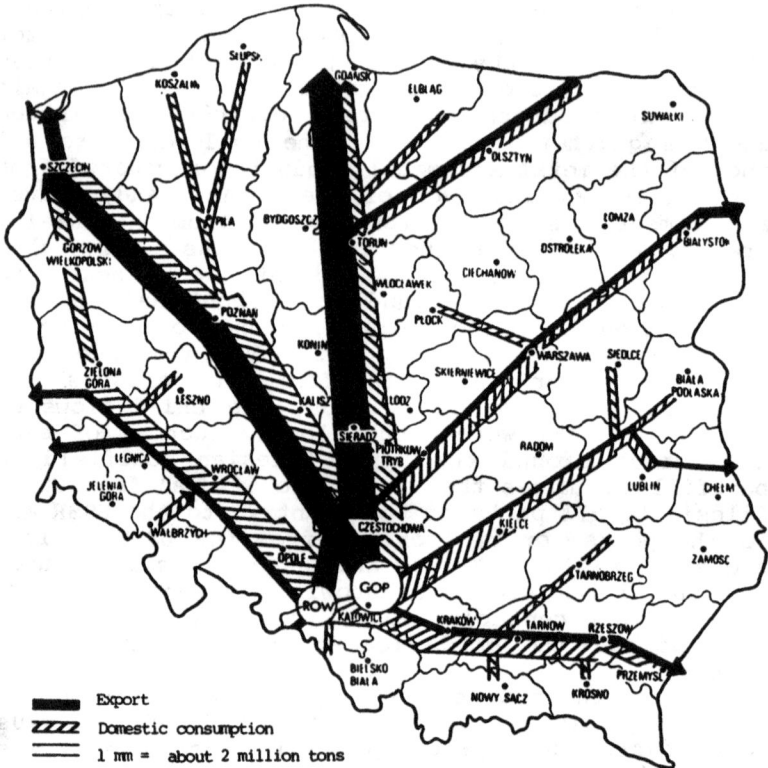

Source: Dziadek, EK 5/1979, revised in
Berman-Alvstam (1980), p.110.

Note: GOP: Gornoslaski Okreg Przemysłowy (Upper
            Silesian Industrial Region)
       ROW: Rybnickie Okreg Weglowy (Rybnik Coal
            District)

of the more general economic crisis in Poland at the
beginning of the 1980s, which gave rise to the birth
of Solidarity.

The Transport Sector in Polish Economic Planning

The Demand Side

The Polish transport crisis of the late 1970s had its roots in both the demand for, and supply of, facilities. On the demand side there were three factors in particular which contributed to it, namely, the adoption of a material and energy-intensive mode of production in the economy in general, allied to the tariff structure which was introduced on the railways; the distribution of the production and consumption of goods across the country, and especially of coal and steel, which had resulted from the spatial development policies which had been adopted; and, thirdly, the effects of the autarkic policies of the Council for Mutual Economic Assistance (CMEA). Each of these will be examined in turn.

The Material and Energy-Intensive Mode of Production

Following the adoption of a policy of rapid industrialisation in the Soviet manner, that is, based upon heavy industries, the output of those industries grew at a rapid rate and without interruption throughout the 1950s and 1960s. Thereafter it was further stimulated by Gierek's policy of financing investment to an increasing extent from Western Europe in return for exports of, amongst other things, coal. In addition, he launched ambitious programmes for the consumer-goods industries and housebuilding. The effect of this growth upon the demand for freight transport was exacerbated by the subordination of the choice of technology which was used in the growing industries to a desire to keep down the cost of investment per unit of output, even where this led to increased demands for inputs of raw materials and energy. As a result, the demand for raw materials rose further, and the capacities of the coal, steel and cement industries had to be expanded yet again. Energy consumption per capita grew rapidly in Poland after the first oil crisis of 1973/74, while it was declining in Western countries. More to the point, seventy per cent of the energy supply in Poland comes from coal, and this coal has to be delivered to hundreds of big consumers, and to about 30,000 smaller ones, distributed all over the country. About 95 per cent of coal transport (in ton-kilometres) is provided by the railways, and, as has been noted above, the carriage of coal, coke, ores and metals accounts for sixty per cent of the total freight on the Polish railways, excluding transit. In the USSR, in contrast, these goods only account for 32 per cent.

The effect of the adoption of a material and energy-intensive mode of production upon the railways has been increased by the pricing system both for the materials themselves and for their transport. Low prices were charged for raw materials during the six-year plan in order to assist the development of heavy industry, and during the 1970s, when world market prices more than quadrupled and domestic mining costs in Poland trebled, prices remained unchanged. In consequence, the cost of coal production at the beginning of the 1980s was estimated to be five to seven times higher than the average sale price (Wołowski, P.T.I. 27/1981). Similarly, freight rates on coal, ore, cement and steel have been held down since the 1950s (Tarski 1968, p.237), with the result that, by the beginning of the 1970s, the marginal costs of increases in the volume of goods carried were above the average costs because of congestion on the railways, and rising faster, while revenues from the bulk transport of goods only covered a small part of their costs on Polish State Railways [2]. What is more, it has proved difficult to alter these prices. Freight rates were not changed between 1960 and 1975, although costs increased two- or three-fold during the same period. Thus, uniform delivered prices were charged for coal within Poland, and so customers who were hundreds of kilometres from the mines were paying the same for transport as those who were next-door to them. In short, coal was not only excessively cheap, but it was especially so for those customers who were far from the coalfields. It is hardly surprising in these circumstances that transport costs have been given very little attention in location decisions in Poland since the war (Kuziemkowski, PK 7, 1977, p.261), or that the industrial location decisions which have been taken have often increased the demand for transport unnecessarily.

The high level of demand for rail transport is also due in part to the phenomenon of irrational transport, which is typical of centrally-planned economies, and which has increased in spite of the shortage of capacity on the railways. In the tariff reforms of 1976 a penalty charge of fifty per cent was introduced on movements which the PKP considered to be unjustified, but the response of the despatchers was to add these charges to their costs and to raise their prices accordingly (P.T.I.,

16/9/1968), and thus the problem has continued.

The Distribution of Production and Consumption across the Country. The demand for transport facilities has also reflected the distribution of economic activity across the country. There are very few countries in which the basic mineral resources and heavy industries are as spatially concentrated as in Poland, and, although the authorities have adopted policies of decentralisation and the development of backward areas, in practice heavy investments have been made in the traditional industrial regions in order to take advantage of the economies of agglomeration and of scale there. In this, Upper Silesia has been particularly favoured, and, during the Gierek period, when the spatial concentration of development was emphasised, Poland's newest steel mill, Huta Katowice, was located at Dąbrowa Górnicza on the north-eastern edge of the area.

As we have seen, Katowice voivodship dominated the pattern of freight movements on the Polish railways, and the further growth of industry within Upper Silesia has increased the problems facing the railways there in a number of ways. Almost all transport to and from the area is by rail, yet the lines pass through heavily populated areas. Most of the lines which are used by coal trains must also be used for other types of freight and passenger transport; and furthermore, an increasing imbalance has been developing between the quantities of goods which are loaded and unloaded. By 1980 the volume of loaded goods was 4.2 times that of unloaded, and the 'surplus' of loaded goods had reached 131 million tons (RST 1981, p.121).

Two alternatives to the increasingly congested railway links within Upper Silesia, and between there and the Baltic ports, from which much coal is exported, might have been the Oder and Vistula rivers. However, projects for the modernisation of the inland waterways have been withdrawn, and the technical standard of river transport is now even lower than it was before the Second World War. There will be no significant increase in the volume of traffic handled on Poland's inland waterways in the foreseeable future.

Autarkic Policies within CMEA. Attention has been drawn already to the early post-war plans for the

231

resumption of Poland's external trade, and for the development of a balanced demand for the transport of such bulky commodities as coal and ore. However, after the formation of CMEA in 1949 the importance of the USSR as a supplier of raw materials increased sharply, and an entirely different geographical structure of transport movements came into existence, which has greatly increased the demands which have been made upon the Polish railways.

As a result of the 1950 Warsaw agreement between the CMEA countries about international goods transport (Soglasheniye o mezhdunarodnom gruzovom soobshcheniye) a common tariff for transit traffic within the bloc was introduced (ETT). This was intended to promote the exchange of goods within CMEA by setting the tariff at a lower level than the lowest domestic rate in any of the countries involved (Gorizontov 1977, p.62), and these arbitrary tariffs have indeed been attractive to the importing countries within CMEA. However, in the case of Poland the system led to a lowering of the tariffs for such traffic as that between the USSR and East Germany, in spite of the fact that PKP tariffs were already very low and covered only a part of the cost of the transport.

At the same time traffic has been stimulated by another agreement between the CMEA countries to the effect that goods shall be charged loco the exporting country. This means that the importing country pays a freight rate which is fixed at fifty per cent of the transport costs from some country outside CMEA which is considered to be a representative source of the commodity in question. For example, the charge for carrying iron ore through the Soviet Union by rail to the Polish border is set at fifty per cent of the rate for sea freight between Swedish and Polish harbours. Thus, during the 1970s, when those rates varied between three and six U.S. dollars per ton, transport over the 1,200 kilometres or so between Krivoy Rog and the Polish frontier cost between 1.5 and three U.S. dollars, converted into transferable rubles.

Similar situations exist with regard to the transport of coal and oil. For instance, the average distance from the Soviet coal mines to the country's borders with Eastern Europe is about 1,200 kilometres, but some pits are much further away (Gorywoda 1978, p.99). There is a large deficit of fuels in the western part of the country which has been met increasingly by deliveries of coal from the Kuzbass or Karaganda to the central and south-western parts of the USSR. In 1979 these deliveries amounted to 30 million tons (or twenty per cent) of the total production in Kuznetsk (Kamenev, Gidrougol, quoted in <u>Coal International</u>, October 1981, p.6). If one takes into account the fact that the coal delivered from the Donbass to Eastern Europe has to be replaced by coal from the Kuzbass, the real distance of transport rises to between 3,000 and 4,000 kilometres. Similarly, in the case of oil, production in the European USSR covers only a part of the local demand, and exports from the Volga-Ural fields to Eastern Europe have to be replaced with oil from the Samotlor and Surgut areas.

Thus, the freight rates in the USSR for exports to Eastern Europe are fixed without regard to either the mode of transport or prime costs, and the fact that they are very low has made it possible for the USSR to establish itself as a main supplier of raw materials (Table 9.3). Of course, Poland has enjoyed these very low freight charges for railway transport on Soviet territory, but it has also been subjected to a similar system of charging on its coal exports to the Soviet Union, and has suffered big losses on the transit traffic, which amounted to 7,000,000 tons in 1980, between the USSR and the GDR. Revenues from this traffic only cover a small fraction of PKP's costs. (Starting in 1984, the freight tariffs between the CMEA countries have been raised considerably. According to the new agreement (MIT) tariffs will be adjusted to world market prices charged for similar transport services [Tymoszuk, PK/8/1983, p.229].) Moreover, Poland has experienced a great increase in the demands upon its rail network arising out of the consequent long-distance commodity movements.

The effect of this dependence upon Soviet raw materials is greatest in the case of the Polish energy and steel sectors. For instance, after the plan for a steel mill to the west of Upper Silesia had been abandoned in 1948, work began at Nowa Huta to the east of Kraków - a site which was chosen in order to facilitate the import of iron ore from the USSR - and since 1949 seventy to eighty per cent of Poland's ore imports have come from that country. The effects of this upon the demand for freight facilities have been several. Firstly, there has been a great increase in the quantity of traffic from the Soviet border to Kraków, Huta Katowice and Upper Silesia, amounting to about 13 million tons per annum. Secondly, there are problems arising out

of the use of different railway gauges in Poland and
the Soviet Union. Enormous investments have been
made in the construction of huge reloading and
axle-changing facilities at Małaszewicze, Medyka and
other places, but all freight which is not in
containers must still be reloaded. In winter coal
and ore arrive frozen, and must be thawed before
this can be done, thus increasing still further the
real costs of transport. However, these extra costs
are not reflected in the tariffs which are charged.
Thirdly, the spatial pattern of traffic has been
unbalanced (Figure 9.4). Within Poland there is an
imbalance between the movements of iron ore and
coal. In 1980 sea-borne imports of iron ore amounted
to only six million tons. Conversely, exports of
coal through the ports amounted to 21 million tons,
while those to the USSR, through Medyka and other
railway crossings, were only three or four million
tons. Some Polish transport economists have reacted
to this situation by suggesting that Huta Katowice
should be supplied with ore from seaborne sources,
rather than from the Soviet Union (Madeyski,
Lissowska and Morawski 1975, pp.21 ff), but their
suggestion was not taken up. More generally, the
imbalance between the USSR and Eastern Europe in the
movement of coal and ore is shown in Figure 9.4,
from which it may be seen that this problem is not
peculiar to Poland. In 1980 Soviet exports by rail
totalled 89 million tons, most of which went to CMEA
countries, while imports were only 19 million
(Vneshnyaya torgovlya SSSR). Most of the exports of
ore to Czechoslovakia, Hungary and Poland probably
came from Krivoy Rog through Znamenka, Kazatin,
Zdolbunov, Lvov and Medyka or occasionally through
Znamenka, Zhmerinka, Ternopol, Mukachevo and Chop.
Smaller quantities are delivered from the Kursk
Magnetic Anomaly, and deliveries to Huta Katowice
via the new LHS railway (see below) are sent from
Zdolbunov via Kovel and Hrubieszów. Finally, by
locating the Lenin steelworks at Nowa Huta, and
later Huta Katowice at Dabrowa Górnicza, further
strain has been imposed upon the hard-pressed
railways through the central parts of Upper Silesia
because metallurgical coal and coke are supplied
from the western part of that coalfield.

Thus, a variety of circumstances has conspired
to encourage long-distance haulage of large
quantities of heavy materials on the Polish railways
since the late 1940s, in a pattern of movements
which is spatially unbalanced, and in return for
receipts which do not reflect the costs of supplying

Figure 9.4
Transport imbalances: Ore and coal deliveries from the USSR to Eastern Europe and return bulk goods traffic in the year 1980. The figure shows railways utilized for outward deliveries of ore and coal via border stations. The main share of the traffic is carried on the electrified lines, but the Kursk (KMA) - Konotop and Zdolbunov - Vladimir Volynskiy via Kovel routes, as well as the Lvov-Vladimir Volynskiy route, are not electrified.

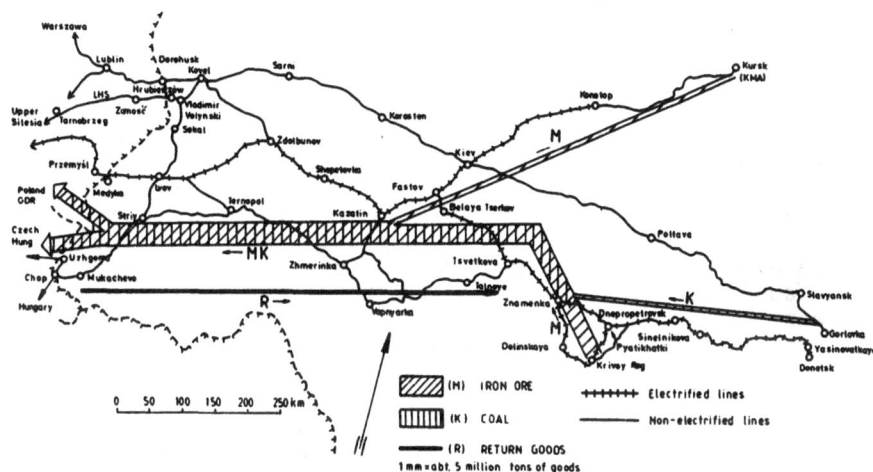

Sources: Ore and coal railway lines according to Galitskiy et al. (1965) pp. 119 and 163; Electrified lines; Nikolskiy (1978), p.122; ore and coal flows: Vneshnyaya torgovlya SSSR (1981), p.15ff.

the transport facilities. What is more, much of this movement pattern is directly the result of the implementation of policies preferred by the Soviet Union with regard to the means of economic development, the spatial distribution of production and the pattern of trade in Eastern Europe.

The Supply Side

The reaction of the Polish authorities to these increasing demands upon the railway system has varied greatly since the late 1940s. We have noted already the trend of railway construction during the 1950s and the beginnings of electrification. Nor should it be forgotten that other forms of transport have also been improved. During the 1960s increased provision was made for the modernisation of the road network, and the Friendship pipeline for oil exports from the USSR to Poland and East Germany was inaugurated. Nevertheless, it was the railways which made the greatest contribution towards carrying the growing freight traffic, at least as it was measured in ton-kilometres, up to 1970, but by 1969-70 PKP was unable to fulfil its planned targets. This failure was due in large part to inadequate investment, and so transport investment was doubled for the 1971-75 five-year plan, and this higher level of expenditure was maintained during the second half of the 1970s. 35 to 40 per cent of this was allocated to the railways, chiefly to increase their freight-carrying capacity (Table 9.4).

The Railway Network. Several new projects were begun during the 1970s with the aid of these investment funds (Figure 9.2), and progress on them up to the end of 1983 was as follows:

(1) The Silesia - Szceczin - Świnoujście coal railway - the aim was to extend and thoroughly modernise the line from Wrocław to Świnoujście via Głogów and Kostrzyń in order to create an electrified railway for coal transport with an annual capacity of forty million tons. 270 kilometres of double track had been completed by 1975, and the electrification is to be completed by 1985.

(2) The main line from Upper Silesia to Gdańsk and Gdynia via Gniezno - the aim is to modernise a series of single-track lines which were used only

for local traffic in order to create a new electrified double-track line to the coast, and thus relieve the pressure on the Katowice-Gdynia line via Bydgoszcz. However, the second track had only been laid between Oleśnica and Gniezno by the end of 1983.

(3) The east-west connection between Hrubieszów, Kielce and Silesia - a normal-gauge main line was planned from the Soviet border in order to relieve the pressure on the Medyka - Kraków - Katowice line, which is particularly used for iron ore imports and coal exports. It was also intended that the new line would improve connections between the south-east of Poland, Częstochowa and Silesia. The Nisko-Zwierzyniec section was completed between 1971 and 1976, but the project was then discontinued after the decision to build the broad gauge LHS railway (see (5) below).

(4) The central main line from Silesia to Gdańsk via Warsaw (C M K) - during the first stage of this project a double-tracked high-speed railway was to be constructed from Katowice to Warsaw through Zawiercie and Grodzisk, and its extension to Gdańsk was to form the second stage. Between 1971 and 1977 one track of 233 kilometres was built from Zawiercie to Grodzisk, but the electrification remains unfinished and the line can only be used for irregular goods traffic. It has, however, been carrying some passenger services since 1984.

(5) The steel and sulphur line - Linja hutniczo-siarkowa (L H S) (Figure 9.2) - in 1976 it was decided to build a broad-gauge railway over the 400 kilometres between Hrubieszów on the Soviet frontier and the new steel mill at Huta Katowice, and the line was completed in 1979 with Soviet assistance. 226 kilometres of the total are completely new, and the rest runs parallel to the pre-existing railway. The line is single track, and awaits electrification. No stations have been built, nor has a modern signalling system been installed. It is not connected to the Polish normal-gauge network, and is therefore no more than a long industrial line for the carriage of iron ore. Moreover, it terminates at Sławków, about ten kilometres from Huta Katowice, and ore must be forwarded by a conveyor belt to the mill. Other steel mills could be supplied by the line, but they too would be obliged to have the ore reloaded at

Slawków. During 1982 5.6 million tons of freight
was carried (PK 3/1983, p.84).
    Thus, much was planned during the 1970s, but,
as several of the projects are unfinished, the
increase in the capacity of the Polish railway
system as a result of the considerable outlay has
only been slight.

Other Equipment. Nor has the network and other
equipment been improved at a sufficient rate in
other respects, as many examples demonstrate. For
instance, although the more heavily-used lines have
been fitted with continuously-welded rails, only
about 20 per cent of the normal-gauge track was of
the heaviest rails (weighing 60 kg/m) by 1980. Or
again, the pace of electrification, which was 350
kilometres per annum between 1971 and 1977, dropped
to 185 between 1978 and 1980, while 2,800 kilometres
of narrow-gauge lines, or as much as in all other
European railway systems together, were still in use
despite the fact that the operating costs of such
lines are two to three times higher than those of
normal gauge (Table 9.5). It should also be noted
that by 1982 centralised traffic control had only
been installed on 3 per cent of the network, and no
attempt had been made to modernise the signalling
system.
    Similar criticisms can be levelled against the
development of goods terminals. Those which handle
larger volumes are well equipped but there is a
great shortage of mechanised facilities in most of
the 2,300. Although 45 specialist coal terminals -
one in almost every voivodship - were supposed to
have been built, no other efforts have been made to
rationalise terminal activities, and this is also
the case in respect of marshalling yards, car and
engine sheds, and repair works. Most are
characterised by out-of-date, neglected or worn-out
equipment.
    Nor is the situation with regard to rolling
stock satisfactory. Before the war Poland
specialised in the production of railway engines and
other rolling stock of a high technical standard.
Engines from Fablok in Chrzanów and Cegielski in
Poznan, as well as cars from Lilpop in Warsaw, were
exported all over the world. But since then it would
appear that insufficient attempts have been made to
uphold this reputation, in spite of the fact that,
as a result of its new frontiers, Poland acquired
several well-known railway works, including Pafawag

238

in Wrocław, Zastal in Zielona Góra, and a car factory at Świdnica. The failure is particularly apparent in the case of diesel and electric engines. During the late 1970s eight-axled, twin-sectioned engines of the Bo-Bo type were delivered to PKP for use with 3,500-ton coal trains. However, domestic production of the engines was insufficient, and extra diesels were imported from the USSR and Romania, while electric engines were bought from Czechoslovakia, East Germany and Great Britain (Table 9.6). Even so, about 4 per cent of freight traffic and 19 of passenger are still hauled by steam engines, and the stock of engines, which increased rapidly during the 1970s, has stagnated since then. A similar failure is apparent in the supply of freight cars. 48,000 were delivered to PKP between 1971 and 1975, and 60,000 between 1976 and 1980, but the annual rate of delivery fell to 2,100 in 1982 (PK 6/1978, p.203; EK 5/1982, p.119) as production of the cars declined by 70 per cent while deliveries of them to the USSR continued. As a result the stock of cars decreased from 197,000 in 1970 to 168,700 in 1982 (Table 9.7), and the situation has been exacerbated by an increase in the number of these which are out of use each day, which rose from 14,500 in 1970 to 59,000 (PK 9/1983, p.263), and by increasing turnround times for them. The shortage of cars has become increasingly obvious since the mid-1970s.

Productivity. The growing problems associated both with major new, and with routine, investment were also being accompanied by disappointing levels of productivity during the late 1970s. For much of the post-war period labour productivity on the railways had been rising, and the number of traffic units per employee rose from 217,000 in 1950 to 497,000 in 1978. More particularly, between 1960 and 1978 the number of ton-kilometres achieved was doubled, and passenger transport increased by 50 per cent. Nevertheless, the railways have become seriously undermanned, with 32,000 vacancies out of an establishment of 410,000 in 1982, and the number of traffic units per employee has fallen sharply since 1978 (Table 9.8 and EK 1/1983, p.17).

Some of the increase in labour productivity was a consequence of the introduction of larger freight cars, and the average waggon load increased from 21 tons in 1965 to 34 in 1982, while the average gross weight of freight trains rose from

1,085 to 1,250 tons. However, average freight train velocity fell from 21.1 kilometres per hour in 1968 to 18.2 in 1979 in spite of a rise in the proportion of the traffic hauled by the faster diesel and electric engines, from 40 to 86 per cent over that period (Nowosielski, PK 5/1981, p.167), while the turnround time for freight cars increased from 4.5 days in 1965 to 5.3 in 1982[3]. By the beginning of the 1980s only 11 per cent of a car's time was spent in motion, and only 7 per cent in loaded motion (EK 7/1981, p.229). The reasons for the declining productivity of the stock are various, but frequent disruptions of traffic occasioned by congestion, low standards of maintenance, and declining labour discipline, which manifests itself in a lack of care in loading and unloading, may all have played a part. If there had been a return in 1982 to the 1965 turnround time for cars, the number which would have been required would have been 25 to 30,000 fewer than it was.

An Assessment of the Investment Programme of the 1970s. Thus, there were a large number of areas in which the Polish railways failed to meet the requirements which were set for them by the ambitious targets for economic growth during the 1970s, and these failures occurred in spite of a substantial increase in investment in the transport sector. Why should this have happened? One answer lies in the balance of the investment. While it is true that there was an increase in investment in transport, it is a fact that there was an even greater increase in investment in industry and construction, and the quantity of investment in the railways was probably insufficient. However, there was also a clear failure to translate the increased spending into additional system capacity, resulting from a mistaken order of priorities in the projects which were undertaken and a failure to complete those which had been begun. Resources were concentrated upon the construction of new lines which would have been better used to improve the existing network, increase the rate of electrification, extend the centralised system of traffic control or modernise the freight terminals. In particular, the building of the LHS line was at the expense of improved north-south connections and the electrification programme (PK 4/1981, p.147). Although that line was intended to be the largest transport investment in the history of Poland, it

became the largest misinvestment, for the demand for
Soviet iron ore at Huta Katowice will probably not
exceed four or five million tons per annum, whereas
the planned capacity of the LHS line was originally
30 million tons. Similarly, a lack of continuity in
investment has vitiated the programme for
containerisation. Billions of złoties have been
spent, but the results, other than in the maritime
sector, have been small. For it to be a success such
a system requires the cooperation of all the
participating bodies which might be connected with
the transport of containers, but in Poland there
appears to have been a shortage of suitable handling
equipment at factory level, thus obstructing the
entire operation (PK 3/1982, p.73).

The Future
As a result of all these and other problems Poland's
national income declined by 25 per cent between 1978
and 1982, and industrial production by 10 per cent.
This decline relieved the stress upon the railways,
at least temporarily, but the demand for rail
transport rose again during 1983 as industrial
production recovered, and in that year PKP carried
415 million tons of freight. Nevertheless, another
thirty million had to be rejected because of a
shortage of capacity.
    Recent projections suggest that the total of
freight traffic on all forms of carriers will
decline from the peak of 2,700 million tons in 1980
to 2,100 in 1990, as the weight of goods produced in
the economy as a whole falls from 814 to 780 million
tons (EK 12/1982). Thus, earlier forecasts that the
demand for rail transport would increase from 489
million tons in 1978 to 760 in 1990 (PK 6/1979,
p.201, EK 12/1982, p. 333) may now be seen to have
been unrealistic, and the most difficult problem
awaiting solution therefore is not the extension
of the railway network, but rather to ensure that
the investment which is required to enable the
system to return to its 1978 output is available.
However, the restrictions put on investments during
the early 1980s hit the transport sector very hard.
Its share of the total investment fund amounted to
6 per cent in 1982-83, compared with 10 per cent
during the first part of the 1970s, despite the fact
that transport investments were given low priority
even at that time. In fixed prices, the capital
inputs in PKP declined from 22 billion złoties in 1978
to 8 billion in 1982 (Table 9.4). Thus PKP had to

interrupt the expansion of the railway network, and the limited resources had to be used primarily for the repair and maintenance of fixed installations, and for an increased rate of electrification which was called for by the severe lack of oil. However, it is doubtful whether the drastically decreased resources will be enough either to counteract a further deterioration of the fixed assets of the PKP, or to realise the ambitious electrification programme of 1,240 km in 1983-85 and 4,200 km during the second part of the 1980s (EK 8/1983, p.204). At present it seems that the originally planned, and very urgent, modernisation of out-of-date and worn-out terminals, depots, garages and signal equipment will be postponed until next decade. However, in view of the economic crisis of the early 1980s it is not realistic to envisage any marked increase in the transport investment fund during the 1980s.

Several indicators point towards an increasing share of sea-borne activity in Poland's foreign trade, and a strong demand for the transport of heavy raw materials in the north-south axis in the long run. Special priority should be given to improve the north-south connection and to the rebuilding of the railway network between Upper Silesia and Gdańsk for heavy traffic in order to diminish the transport costs on low-valued goods. On the other hand, because of the long distances of iron ore and coal transport between Poland and the USSR, as well as the fact that these are on lines which are used both for goods and passenger transport, and with existing transport imbalances and the different gauges, it is not realistic to forecast any increased volume of mass-goods movement in the east-west axis. The USSR has encountered increasing problems in the supplying of Eastern Europe with raw materials - iron ore, coal, oil, etc. In a paper presented by Soviet specialists to the XXIII conference of the CMEA countries at Bucharest in September 1982, it was noted that 70 per cent of the USSR exports to the CMEA consisted of raw materials, while 77 per cent of the imports were made up of manufactured products:

> The possibilities of solving the energy and raw material problems of the smaller CMEA countries through increased imports from the USSR are exhausted.... Different objective factors are limiting the production and export possibilities of Soviet fuel and other raw

materials, and a cooperative effort is required by all CMEA countries to keep the export volumes at the present level. This implies that future deliveries to a certain CMEA country, even at an <u>unaltered</u> (our underlining) level is dependent on the participation of that country in Soviet investments in excavation industries or in branches of industry which serve the excavation sector (Danielewski, Z.G. 45/1982).

Accordingly, the Soviet Union has obliged Eastern European countries to join new projects for the exploitation of mineral resources and to provide financial support for them, or to purchase machinery and equipment for Soviet mines from the West using hard currency. Such projects have become an increasingly heavy burden upon the Polish economy in a period in which the Polish investment fúnd has been in decline, and therefore circumstances for both the USSR and Poland would seem to indicate that a new policy is required which will encourage the purchase of raw materials on the world market by the Poles. However, the shortage of convertible currencies and the political relations between the superpowers may yet prove to be the enemy of the efficient development of the Polish space economy and its railway network, for deteriorating East-West relations in the early 1980s led to a reaffirmation of the policy of self-sufficiency at the CMEA meeting in Moscow June 1984. It would be unfortunate if the logic of Poland's economic geography, which was set aside in the six-year plan, should fall prey again to the dictates of the country's political location.

Table 9.1   Total domestic freight traffic, by mode, 1950-1983

Year	Total	Railways	Roads	Inland waterways	Pipeline	Other	PKP share(%)
				Million tons			
1950	351	160	110	1	–	80	46
1960	714	287	408	3	–	16	40
1970	1272	382	863	9	15	3	30
1975	2254	464	1743	15	31	1	20
1978	2662	489	2110	22	41	–	18
1980	2713	482	2168	22	41	–	18
1981	2031	402	1576	17	36	–	20
1982	1832	402	1379	14	37	–	22
1983	1865	415	1397	14	39	–	22
				Thousand million ton-kilometres			
1950	37	35	1.4	0.3	–	0.3	95
1960	73	67	5.7	0.9	–	–	91
1970	124	99	16	2.3	7.0	–	80
1975	176	129	33	1.9	13	–	73
1978	200	138	43	2.7	17	–	69
1980	199	135	45	2.3	17	–	68
1981	165	110	37	1.9	16	–	67
1982	165	113	34	1.6	17	–	68
1983	172	118	35	1.5	17	–	68

Sources: 1950-1975: RST 1976, pp.xx-xxiii;
1978-1982: RS 1983, p.317;
1983: MRS 1984, p.202

Table 9.2 Polish State Railways: Goods transport by commodity (excluding transit), 1960-1982

Commodity	1960	1970	1980	1982	1960	1970	1980	1982
	Thousand million ton-kms				Percentage composition			
TOTAL	57.7	89.5	124.8	104.3	100	100	100	100
Coal and coke	24.4	36.4	51.6	49.1	42	41	41	47
Ores	3.7	4.0	9.4	6.2	6	6	8	6
Metals	3.0	7.2	10.1	7.3	5	8	8	7
Oil and petroleum products	1.3	3.4	5.7	4.3	2	4	5	4
Bricks, stone, sand, gravel	7.4	7.7	10.9	6.8	13	9	9	7
Cement	1.5	2.5	3.3	2.9	3	3	3	3
Fertilizers, other chemicals	2.3	6.9	9.7	9.1	4	8	8	9
Grain, other agric. products	4.0	4.6	5.9	4.0	7	5	5	4
Timber	3.2	3.6	3.6	2.9	6	4	3	3
Other goods	6.9	12.3	14.6	11.7	12	12	10	10

Sources: 1960-1970: RST 1971, p.31f;
1980-1982: RST 1983, p.317f:

Table 9.3 Iron ore exports from the Soviet Union to CMEA countries 1955–1982 (million tons)

Country	1955	1960	1970	1975	1980	1981	1982
Bulgaria	-	-	1.0	1.7	2.2	2.3	2.3
Czechoslovakia	3.0	5.1	10.8	12.2	10.3	9.7	9.9
GDR	1.2	2.0	2.7	2.7	3.2	3.1	2.8
Hungary	1.3	1.7	3.9	4.0	3.1	3.4	3.6
Poland	3.1	5.2	9.9	11.1	13.7	10.9	9.0
Romania	0.2	0.9	4.2	6.3	4.2	4.3	4.4
CMEA total	8.8	14.9	32.5	38.0	36.7	33.7	32.0
Other countries	-	0.3	3.6	5.6	1.4	1.2	1.2
TOTAL	8.8	15.2	36.1	43.6	38.1	34.9	33.2

Source: Vneshnyaya torgovlya SSSR

Table 9.4   Transport investment in Poland by mode, 1970-1982

Year	1977 prices		Current prices (000 million zl)					The share of PKP in total transport investment (%)
	Total fund (bill zl)	Transport (%)	Total	Railways	Road Transport	Shipping	Others	
1970	279	11.0	24.0	8.9	6.1	2.9	6.1	37
1971	299	11.8	29.8	11.2	6.8	4.9	6.9	38
1972	368	10.3	31.8	12.4	7.5	3.4	7.5	39
1973	462	9.3	35.7	12.8	9.5	4.1	9.3	36
1974	564	10.2	48.1	16.9	10.3	8.7	12.2	35
1975	625	10.3	57.8	19.8	12.7	12.9	12.4	34
1976	631	8.4	52.4	20.1	14.5	5.4	12.4	38
1977	651	7.7	50.2	21.6	14.8	3.1	10.7	43
1978	665	8.3	58.2	24.1	15.6	5.0	13.5	41
1979	612	8.2	53.0	21.3	16.3	4.6	10.8	40
1980	537	9.6	56.1	19.5	20.0	4.4	12.2	35
1981	415	7.4	34.7	13.3	12.5	1.0	7.9	38
1982	365	5.9	56.3	21.3	16.4	2.5	16.1	38

Sources : 1970-1974: Our estimates from RS 1982, pp.XXX and XL
1975-80: RS 1982, p.143 and RST 1981, p.XXX; 1981: RS 1982, p.143.
1982: converted to 1977 prices according to RS 1983, p.156.

Table 9.5 PKP's network length, 1950–1983

	1950	1960	1970	1975	1980	1983
Length of line operated (km)	26312	26904	26678	26702	27185	27176
Normal gauge	22482	23232	23311	23766	24356	24366
of which:						
– electrified	156	1026	3872	5588	6868	7828
– single track	16333	15977	15713	15488	15599	..
Percentage of tracks equipped with *)						
– 60 kg/m rails	–	–	(1)	9.9	20.2	..
– long welded rails	–	–	(10)	20.4	36.2	..
Narrow gauge	3830	3672	3367	2936	2829	2810

Sources: 1950–70: RST 1976, p.28;
1975–80: RST 1981, p.4f;
1983: MRS 1984, p.199.

*) Percentage of main running tracks excluding sidings, etc.
The 1970 figures are estimated.

Table 9.6 Rolling stock production and trade, 1950–1983

	1950	1960	1970	1975	1980	1982	1983
Production							
Diesel engines	–	144	351	421	121	30	41
Electric engines	–	38	75	75	125	91	77
Electric railmotor vehicles	–	23	44	62	76	50	50
Passenger cars	158	660	569	543	328	308	350
Freight-carrying cars	13900	13900	15500	18700	15200	6400	6000
Import							
Engines	–	65	169	146	247	78	62
Freight-carrying cars	–	596	222	1644	–	–	–
Export							
Freight-carrying cars	4006	3932	5779	6924	6745	4324	3350

Sources: 1950–1970: RS 1974, p.217; RSHZ 1975, p.30ff;
1975–1982: RS 1983, pp.195, 341;
1983: MRS 1984.

Table 9.7   Polish State Railways' working fleets, 1970-1982

	1970	1975	1978	1980	1982	1983
Electric engines	665	1047	1262	1406	1486	1572
Diesel engines	947	1893	2638	2832	2516	2556
Steam engines	3462	2582	1806	1442	1132	1043
Electric railmotor vehicles	..	705	795	887	..	..
Diesel railmotor vehicles	..	205	168	151	..	..
Passenger cars	..	6178	6038	5734	..	..
Freight-carrying cars, total (000)	197.3	212.0	211.3	194.9	168.7	172.2
of which:						
- covered	48.6	45.0	37.7	33.3	29.9	31.3
- coal cars	112.0	120.7	119.5	115.2	97.2	97.6
- flat	21.6	21.1	22.1	21.5	19.3	23.2
- special wagons	15.1	25.2	32.1	24.9	22.3	20.1
Share of four-axle units (%)	17.4	32.4	40.1	45.3	46.0	..
Average age (years)	24.0	17.0	16.5	16.3	15.5	..
Average carrying capacity (tons)	27.5*)	..	..	..	40.3	..

Sources: RST 1981, p.14; RS 1983, p.318; EK 8/1983, p.207; MRS 1984, p.204.
*) 1971

Table 9.8. Railway employment in Poland, traffic and labour productivity, 1960-1982

Year	Employed (thousands)	Ton-kilometres (000 million)	Passenger-kms (000 million)	Traffic units (000 million)	Traffic units per employee (thousands)
1950	286	35.1	27.1	62.2	217
1960	350	66.5	30.9	97.4	278
1970	361	99.3	36.9	136.2	377
1978	372	138.1	46.7	184.8	497
1980	372	134.7	46.3	181.0	481
1982	375	112.5	49.2	162.0	432

Sources: 1950-1975: RST 1976, p.XXf;
1978-1982: RS 1983, p.313ff.

NOTES

[1] According to the Instytut Transportu Samochodowego (Institute for Road Transport) the Central Bureau of Statistics (GUS) overestimated the road transport performance by 15-20 per cent during the 1960s (Malek, 1973, p.37). This overrating is now estimated to be in the order of 10-15 per cent (Kuziemkowski, PK 5/1981, p.142), which means that the share of the railway is higher than that quoted in Table 9.1.

[2] Low tariffs for low-value commodities were also charged in market economies after the Second World War, for example for coal in Britain. The chief reason was to increase traffic on underused transport systems, and thus make full use of existing infrastructure. In Poland, in contrast, in the 1970s, the difference between the tariffs on such commodities and those on industrial products was much greater than it had been in Western European countries, and that at a time when the Polish railway system was already congested with goods traffic.

[3] It should be noted that PKP counts the car turnround time in relation to the working fleet in operation, while the EEC uses the total working fleet including unusable cars. According to the EEC definition the Polish car turnround time for 1982, with 169,000 units in operation and 59,000 out of use, was as high as 7 days.

REFERENCES

Berezowski, S. (ed) (1983) Geografia ekonomiczna Polski, PWN, Warszawa (Second ed. 1978)

Berman, Z. - Alvstam, C.G. (1980) Transporter i Polen, del III: Gods pa järnvägarna, Choros, Göteborgs Universitet, Göteborg

Berman, Z. - Alvstam, C.G. (1983) Polsk kol-och stalindustri, energifor-sörjning och transportstruktur, Choros, Göteborgs Universitet, Göteborg

Galitskiy, M.I. et al (1965) Ekonomicheskaya geografiya transporta SSSR, Moskva

Główny Urząd Statystyczny (1981), Wojewódzkie Bilanse Przewozów Ładunków Przez PKP i Żeglugę Sródlądową 1981, Warszawa

Gorywoda, M. (1978) Wsólpraca krajów RWPG w gospodarowaniu surowcami, Warszawa

Gorisontov, B.B. (1977) Sozialistische ekonomische Integration und Transportwesen, Transpress, Berlin

Kusmierek, J. (1980) quoted in The Times, London 31 October 1984

Lijewski, T. (1977) Geografia transportu Polski, PWE, Warszawa

Madeyski, M. - Lissowska, E. - Morawski, W. (1975) Transport - rozwój i integracja, WKL, Warszawa

Malek, P. et al (1973) Ekonomica transportu samochodowego, WKL, Warszawa

Malisz, B. (1974) Problematyka przestrzennego zagospodarowania Kraju, P.W.N., Warszawa

Nikol'skiy, I.V. (1978) Ekonomicheskaya geografiya transporta SSSR, Moskva

Tarski, I. (1968) Koordynacja transportu PWE, Warszawa

Wróbel, A. and Zawadzki, S.M. (1966) Location Policy and the Regional Efficiency of Investments in City and Regional Planning in Poland, ed. J.C. Fisher, Cornell, Ithaca, pp.433-40

Newspapers and Journals
Coal International
Eksploatacja Kolei (EK) Warszawa
Handel Zagraniczny (HZ) Warszawa
Przegląd Komunikacyjny (PK) Warszawa
Przegląd Techniczny-Innowacje (PTI), Warszawa
Rail International, Bruxelles
Technika i Gospodarka Morska (TGM), Warszawa
Życie Gospodarcze (ZG), Warszawa

Statistical sources
Rocznik Statystyczny (RS), Warszawa
Rocznik Statystyczny Gospodarki Morskiej (RSGM), Warszawa
Rocznik Statystyczny Handlu Zagranicznego (RSHZ), Warszawa
Rocznik Statystyczny Transportu (RST), Warszawa
Vneshnyaya Torgovlya SSSR, Moskva

LIST OF CONTRIBUTORS

Claes G. Alvstam, Kulturgeografiska Institutionen, University of Gothenburg, Sweden

John Ambler, associated with members of the Centre for Russian and East European Studies, University of Birmingham; Management Consultant (PA Consulting Services), England

Zygmunt Berman, Kulturgeografiska Institutionen, University of Gothenburg, Sweden

Martin Crouch, Department of Politics, University of Bristol, England

Andrew H. Dawson, Department of Geography, The University, St. Andrews, Fife, Scotland

Derek R. Hall, Department of Geography and History, Sunderland Polytechnic, England

Holland Hunter, Department of Economics, Haverford College, Haverford, Pennsylvania, U.S.A.

Paul E. Lydolph, Department of Geography, University of Wisconsin-Milwaukee, Milwaukee, Wisconsin, U.S.A

Denis J.B. Shaw, Department of Geography, University of Birmingham, England

Leslie Symons, Department of Geography, University College of Swansea, Wales

Johannes Tismer, Department of Economics, Osteuropa-Institut, Free University of Berlin, Berlin, Federal Republic of Germany

John N. Westwood, Centre for Russian and East European Studies, University of Birmingham, England

David Wilson, School of Geography, University of Leeds, England

254

INDEX

This index is designed to aid the reader in locating all the major topics in this book, many of which are introduced in the Preface. Accordingly, there are no references to personalities and towns, though important regions of the USSR and Poland are listed. The entry 'Anecdotes' may exemplify the method. Communist journals often contain stories - probably true, but not necessarily typical - which are published for various reasons: to demonstrate how socialists can 'do it better', to encourage technique transfer, to show that 'Big Brother' is always there, to reveal new thinking by the leadership, to act as rudimentary market research. All such instances are indexed here under 'Anecdotes'.

Some major themes and key-words are: investment, prices, costs; infrastructure, inner reserves, bottlenecks, spare capacity, rush hour; common carrier transport, private transport, pools; departmentalism, transport policy, planning, modal choice, innovation, regional policy, rural economy; organisation, management, pay, employment; technology, spares, repairs and breakdowns; commodities; COMECON countries; climate, topography.

INDEX

For Product Safety Concerns and Information please contact our EU
representative GPSR@taylorandfrancis.com
Taylor & Francis Verlag GmbH, Kaufingerstraße 24, 80331 München, Germany

www.ingramcontent.com/pod-product-compliance
Lightning Source LLC
Chambersburg PA
CBHW050408280326
41932CB00013BA/1786

* 9 7 8 0 3 6 7 7 2 6 0 9 6 *